CALCULATIONS FOR O-LEVEL CHEMISTRY

E N Ramsden BSc, PhD, DPhil

Wolfreton School, Hull

Stanley Thornes (Publishers) Ltd

Other Chemistry books for schools from Stanley Thornes (Publishers) Ltd

MODERN ORGANIC CHEMISTRY by A Atkinson
A COMPLETE 'O' LEVEL CHEMISTRY by G N Gilmore
A MODERN APPROACH TO COMPREHENSIVE CHEMISTRY by G N Gilmore
REVISION NOTES IN CHEMISTRY by E N Ramsden
A FIRST CHEMISTRY COURSE by E N Ramsden
CALCULATIONS FOR A-LEVEL CHEMISTRY by E N Ramsden

© E N Ramsden 1981

Reprinted 1982

First published in 1981 by Stanley Thornes (Publishers) Ltd,
Educa House, Old Station Drive, off Leckhampton Road,
Cheltenham GL53 0DN

British Library Cataloguing in Publication Data
Ramsden, E. N.
    Calculations for O. Level chemistry.
    1. Chemistry—Mathematics
    I. Title
    540′.1′51   QD39.3.M3

ISBN 0 85950 312 7

Typeset by Quadraset Limited, Radstock, Bath
Printed in Great Britain at The Pitman Press, Bath

Contents

Preface

This book has been written to provide practice in calculations for O-Level Chemistry students. All the types of numerical problems met in GCE O-level syllabuses are covered. A brief treatment of the theoretical background to each type of numerical problem is given, and is followed by a series of worked examples. The problems are divided into three sections of increasing difficulty, the third section being a collection of questions from past O-level papers. The arithmetic in Sections 1 and 2 has been kept simple by basing the problems on a selection of compounds with relative formula masses which are round numbers, such as $NH_4NO_3 = 80$, $MgSO_4 = 120$, $CaBr_2 = 200$. If a pupil has difficulty with a problem, he or she can return for help to the theoretical section and to the worked examples. Thus, the book can be used for private study, as well as for class work.

The concept of the mole is the thread which knits together the calculations on reacting masses of solids, reacting volumes of gases, empirical formulae, volumetric analysis, electrolysis and heats of reaction. The pupil learns to look at the equation and ask himself or herself how many moles of reactant are involved.

The Association for Science Education publication, *Chemical Nomenclature, Symbols and Terminology in School Science* (2nd edn, 1979), which embodies the latest recommendations of the International Union of Pure and Applied Chemistry, is followed in matters of terminology. Concentrations are expressed as the number of moles of solute per cubic decimetre of solution (mol dm^{-3}). The older term, 'molarity', is also explained as it continues to be used in schools and in some examination papers. In addition to the units recommended by IUPAC and ASE, millimetres of mercury and degrees Celsius are used for pressure and temperature. Reference is made to the obsolescent unit of charge, the Faraday.

Some Examination Boards use terms which differ from those recommended by IUPAC, and students should familiarise themselves with the terminology used by their own Board. Reference to the questions from past papers will assist them in this.

Acknowledgements

I thank the following Examination Boards for permission to print questions from recent examination papers. In a few cases, the wording of the questions has been changed slightly, for example, to ask readers to plot results on a piece of graph paper, instead of on the question paper. The Boards are not responsible for the accuracy of the numerical answers.

The Associated Examining Board (at whose request the terminology of some questions has been updated)
University of Cambridge Schools Local Examinations Syndicate
Joint Matriculation Board
Oxford and Cambridge Schools Examination Board
Oxford Delegacy of Local Examinations
Southern Universities' Joint Board
University of London School Examinations Council
Welsh Joint Education Committee

My thanks are also offered to those who have helped me during the preparation of this book, including the pupils who have read the text and given me the benefit of their comments. I am grateful to Chris Baker for helpful discussions and to Stephanie Cox for checking the numerical answers. I thank Stanley Thornes (Publishers) for the care they have taken over the preparation of the text. Finally, I thank my family for their support and encouragement.

E N Ramsden
Hull, 1981

1. Formulae and Equations

Calculations are based on formulae and on equations. In order to tackle the calculations in this book, you will have to be quite sure you can work out the formulae of compounds correctly, and that you can balance equations. This section is a revision of work on formulae and equations.

Formulae

Electrovalent compounds consist of oppositely charged ions. The compound formed is neutral because the charge on the positive ion (or ions) is equal to the charge on the negative ion (or ions). In sodium chloride, $NaCl$, one sodium ion, Na^+, is balanced in charge by one chloride ion, Cl^-.

This is how the formulae of electrovalent compounds can be worked out

Compound	*Zinc chloride*
Ions present are	Zn^{2+} and Cl^-
Now balance the charges	One Zn^{2+} ion needs two Cl^- ions
Ions needed are	Zn^{2+} and $2Cl^-$
The formula is	$ZnCl_2$
Compound	*Sodium sulphate*
Ions present are	Na^+ and SO_4^{2-}
Now balance the charges	Two Na^+ balance one SO_4^{2-}
Ions needed are	$2Na^+$ and SO_4^{2-}
The formula is	Na_2SO_4
Compound	*Aluminium sulphate*
Ions present are	Al^{3+} and SO_4^{2-}
Now balance the charges	Two Al^{3+} balance three SO_4^{2-}
Ions needed are	$2Al^{3+}$ and $3SO_4^{2-}$
The formula is	$Al_2(SO_4)_3$
Compound	*Iron(II) sulphate*
Ions present are	Fe^{2+} and SO_4^{2-}
Now balance the charges	One Fe^{2+} balances one SO_4^{2-}
Ions needed are	Fe^{2+} and SO_4^{2-}
The formula is	$FeSO_4$

Compound	Iron(III) sulphate
Ions present are	Fe^{3+} and SO_4^{2-}
Now balance the charges	Two Fe^{3+} balance three SO_4^{2-}
Ions needed are	$2Fe^{3+}$ and $3SO_4^{2-}$
The formula is	$Fe_2(SO_4)_3$

You need to know the charges of the ions in Table 1.1. Then you can work out the formula of any electrovalent compound.

You will notice that the compounds of iron are named iron(II) sulphate and iron(III) sulphate to show which of its valencies iron is using in the compound. This is always done with the compounds of elements of variable valency.

Table 1.1   *Symbols and valencies of common ions*

Name	Symbol	Valency	Name	Formula	Valency
Hydrogen	H^+	1	Hydroxide	OH^-	1
Ammonium	NH_4^+	1	Nitrate	NO_3^-	1
Potassium	K^+	1	Chloride	Cl^-	1
Sodium	Na^+	1	Bromide	Br^-	1
Silver	Ag^+	1	Iodide	I^-	1
Copper(I)	Cu^+	1	Hydrogen-carbonate	HCO_3^-	1
Barium	Ba^{2+}	2	Oxide	O^{2-}	2
Calcium	Ca^{2+}	2	Sulphide	S^{2-}	2
Copper(II)	Cu^{2+}	2	Sulphite	SO_3^{2-}	2
Iron(II)	Fe^{2+}	2	Sulphate	SO_4^{2-}	2
Lead	Pb^{2+}	2	Carbonate	CO_3^{2-}	2
Magnesium	Mg^{2+}	2			
Zinc	Zn^{2+}	2			
Aluminium	Al^{3+}	3	Phosphate	PO_4^{3-}	3
Iron(III)	Fe^{3+}	3			

The formulae of covalent compounds

To work out the formulae of covalent compounds, you need to know the symbols and the valencies of the elements present. These are listed in Table 1.2. The valency is the number of electrons which an element shares in forming a compound. The method of working out the formulae is the same as for electrovalent compounds, although here electrons are shared, not given and accepted.

Table 1.2 *Symbols and valencies of some common elements*

Element	Symbol	Valency	Element	Symbol	Valency
Bromine	Br	1	Iodine	I	1
Carbon	C	4	Nitrogen	N	3 and 5
Chlorine	Cl	1	Oxygen	O	2
Fluorine	F	1	Phosphorus	P	3 and 5
Hydrogen	H	1	Sulphur	S	2, 4 and 6

Method of working out the formulae of covalent compounds

Compound	*Compound of carbon and hydrogen*
Symbols of elements	C H
Valencies (no. of shared electrons)	4 1
Balance the electrons	One C with four e⁻ needs four H with one e⁻
Atoms needed	C and 4 H
The formula is	CH_4
	You will recognise the formula of methane.

Compound	*Compound of sulphur and hydrogen*
Symbols of elements	S H
Valencies (no. of shared electrons)	2 1
Balance the electrons	One S with two e⁻ needs two H with one e⁻
Atoms needed	S and 2 H
The formula is	H_2S
	This is the formula of hydrogen sulphide.

Equations

Having symbols for elements and formulae for compounds gives us a way of representing chemical reactions.

Example 1 Instead of writing, 'Copper carbonate dissociates into copper oxide and carbon dioxide', we can write

$$CuCO_3 \rightarrow CuO + CO_2$$

The atoms we finish with are the same in number and kind as the atoms we start with. We start with one atom of copper, one atom of carbon and three atoms of oxygen, and we finish with the same. This makes the two sides of the expression equal, and we call it an *equation.* A simple way of conveying a lot more information is to include *state symbols* in the equation. These are (s) = solid, (l) = liquid, (g) = gas, (aq) = in solution in water. The equation

$$CuCO_3(s) \rightarrow CuO(s) + CO_2(g)$$

tells you that solid copper carbonate dissociates to form solid copper oxide and the gas carbon dioxide.

Example 2 The equation

$$Zn(s) + H_2SO_4(aq) \rightarrow ZnSO_4(aq) + H_2(g)$$

tells you that solid zinc reacts with a solution of sulphuric acid to give a solution of zinc sulphate and hydrogen gas. Hydrogen is written as H_2, since each molecule of hydrogen gas contains two atoms.

Example 3 Sodium carbonate reacts with dilute hydrochloric acid to give carbon dioxide and a solution of sodium chloride. The equation could be

$$Na_2CO_3(s) + HCl(aq) \rightarrow CO_2(g) + NaCl(aq) + H_2O(l)$$

but, when you add up the atoms on the right, you find that they are not equal to the atoms on the left. The equation is not 'balanced', so the next step is to balance it. Multiplying NaCl by two gives

$$Na_2CO_3(s) + HCl(aq) \rightarrow CO_2(g) + 2NaCl(aq) + H_2O(l)$$

This makes the number of sodium atoms on the right-hand side equal to the number on the left-hand side. But there are two chlorine atoms on the right-hand side, therefore the HCl must be multiplied by two:

$$Na_2CO_3(s) + 2HCl(aq) \rightarrow CO_2(g) + 2NaCl(aq) + H_2O(l)$$

The equation is now balanced.

When you are balancing a chemical equation, the only way you do it is to multiply the number of atoms or molecules. You never try to alter a formula. In the above example, you got two chlorine atoms by multiplying HCl by 2, not by altering the formula to HCl_2, which does not exist.

The steps in writing an equation are

1. write a word equation
2. put in the symbols and formulae
3. balance the equation

Example 4 The reaction between sodium and water to form hydrogen and sodium hydroxide solution. Work through the three steps:

1. Sodium + water $\rightarrow$ Hydrogen + Sodium hydroxide solution

2. $Na(s) + H_2O(l) \rightarrow H_2(g) + NaOH(aq)$

3. $2Na(s) + 2H_2O(l) \rightarrow H_2(g) + 2NaOH(aq)$

Example 5 Methane burns to form carbon dioxide and water:

$$CH_4(g) + O_2(g) \rightarrow CO_2(g) + H_2O(g)$$

There is one carbon atom on the left-hand side and one carbon atom on the right-hand side. There are four hydrogen atoms on the left-hand side, and therefore we need to put four hydrogen atoms on the right-hand side. Putting $2H_2O$ on the right-hand side will accomplish this:

$$CH_4(g) + O_2(g) \rightarrow CO_2(g) + 2H_2O(g)$$

There is one molecule of O_2 on the left-hand side and four O atoms on the right-hand side. We can make the two sides equal by putting $2O_2$ on the left-hand side:

$$CH_4(g) + 2O_2(g) \rightarrow CO_2(g) + 2H_2O(g)$$

This is a balanced equation. The numbers of atoms of carbon, hydrogen and oxygen on the left-hand side are equal to the numbers of atoms of carbon, hydrogen and oxygen on the right-hand side.

Example 6 Propane also burns to form carbon dioxide and water:

$$C_3H_8(g) + O_2(g) \rightarrow CO_2(g) + H_2O(g)$$

As there are three C atoms on the left-hand side, there must be $3CO_2$ molecules on the right-hand side:

$$C_3H_8(g) + O_2(g) \rightarrow 3CO_2(g) + H_2O(g)$$

As there are eight H atoms on the left-hand side, there must be $4H_2O$ on the right-hand side:

$$C_3H_8(g) + O_2(g) \rightarrow 3CO_2(g) + 4H_2O(g)$$

Counting the oxygen atoms, there are two on the left-hand side and ten on the right-hand side. Putting $5O_2$ on the left-hand side will make the two sides equal:

$$C_3H_8(g) + 5O_2(g) \rightarrow 3CO_2(g) + 4H_2O(g)$$

This is a balanced equation.

Practice with equations

1. For practice, try writing the equations for the reactions:
   (a) Hydrogen + Copper oxide → Copper + Water
   (b) Carbon + Carbon dioxide → Carbon monoxide
   (c) Carbon + Oxygen → Carbon dioxide
   (d) Magnesium + Sulphuric acid → Hydrogen + Magnesium sulphate
   (e) Copper + Chlorine → Copper(II) chloride

2. Now try writing balanced equations for the reactions:
   (a) Calcium + Water → Hydrogen + Calcium hydroxide solution
   (b) Copper + Oxygen → Copper(II) oxide
   (c) Sodium + Oxygen → Sodium oxide
   (d) Iron + Hydrochloric acid → Iron(II) chloride solution
   (e) Iron + Chlorine → Iron(III) chloride

3. Balance these equations:
   (a) $Na_2O(s) + H_2O(l) \rightarrow NaOH(aq)$
   (b) $KClO_3(s) \rightarrow KCl(s) + O_2(g)$
   (c) $H_2O_2(aq) \rightarrow H_2O(l) + O_2(g)$
   (d) $Fe(s) + O_2(g) \rightarrow Fe_3O_4(s)$
   (e) $Mg(s) + N_2(g) \rightarrow Mg_3N_2(s)$
   (f) $NH_3(g) + O_2(g) \rightarrow N_2(g) + H_2O(g)$
   (g) $Fe(s) + H_2O(g) \rightarrow Fe_3O_4(s) + H_2(g)$
   (h) $H_2S(g) + O_2(g) \rightarrow H_2O(g) + SO_2(g)$
   (i) $H_2S(g) + SO_2(g) \rightarrow H_2O(l) + S(s)$

2. Relative Atomic Mass; Relative Molecular Mass and Relative Formula Mass; Percentage Composition

Relative Atomic Mass

Atoms are tiny: one atom of hydrogen has a mass of 1.66×10^{-24}g; one atom of carbon has a mass of 1.99×10^{-23}g. Numbers as small as this are awkward to handle, and, instead of the actual masses, we use relative atomic masses. Since hydrogen atoms are the smallest of all atoms, one atom of hydrogen was taken as the mass with which all other atoms would be compared. Then,

$$\text{Relative atomic mass} = \frac{\text{Mass of one atom of the element}}{\text{Mass of one atom of hydrogen}}$$

Thus, on this scale, the relative atomic mass of hydrogen is 1, and, since one atom of carbon is 12 times as heavy as one atom of hydrogen, the relative atomic mass of carbon is 12.

The modern method of finding relative atomic masses is to use an instrument called a mass spectrometer. The most accurate measurements are made with volatile compounds of carbon, and it was therefore convenient to change the standard of reference to carbon. There are three isotopes of carbon. Isotopes are forms of an element which have the same number of protons and electrons but have different numbers of neutrons, and therefore different masses. It was decided to use the most plentiful carbon isotope, carbon-12. Thus,

$$\text{Relative atomic mass} = \frac{\text{Mass of one atom of an element}}{(1/12) \text{ Mass of one atom of carbon-12}}$$

On this scale, carbon-12 has a relative atomic mass of 12, and hydrogen has a relative atomic mass 1.007 97. Since relative atomic masses are ratios of two masses, they have no units. As this value for hydrogen is very close to one, the value of H = 1 is used in most calculations. A table of approximate relative atomic masses is given on page 145.

Relative Molecular Mass

A molecule consists of a combination of atoms. You can find the mass of a molecule by adding up the masses of all the atoms in it. You can find the relative molecular mass of a compound by adding the relative atomic masses of all the atoms in a molecule of the compound. For example, you can work out the relative molecular mass of carbon dioxide as follows:

The formula is CO_2

1 atom of C, relative atomic mass   12 = 12

2 atoms of O, relative atomic mass   16 = 32

Total   = 44

Relative molecular mass of CO_2 = 44

Relative Formula Mass

There are, however, a vast number of compounds which consist of ions, not molecules. The compound sodium chloride, for example, consists of sodium ions and chloride ions. One cannot correctly refer to a 'molecule of sodium chloride' or a 'molecule of copper sulphate'. For ionic compounds, the term *formula unit* is used to describe the ions which make up the compound. A formula unit of sodium chloride is NaCl. A formula unit of copper sulphate is $CuSO_4$. A formula unit of copper sulphate-5-water is $CuSO_4 \cdot 5H_2O$. The term relative formula mass is used for ionic compounds:

$$\text{Relative formula mass} = \frac{\text{Mass of one formula unit}}{(1/12) \text{ Mass of one atom of carbon-12}}$$

Thus, for sodium chloride:

The formula is NaCl

1 atom of Na, relative atomic mass 23   = 23

1 atom of Cl, relative atomic mass   35.5 = 35.5

Total   = 58.5

Relative formula mass of NaCl   = 58.5

We work out the relative formula mass of calcium chloride as follows:

The formula is $CaCl_2$

1 atom of Ca, relative atomic mass   40   =   40

2 atoms of Cl, relative atomic mass   35.5 =   71

Total $= 111$

Relative formula mass of $CaCl_2$ $= 111$

We work out the relative formula mass of aluminium sulphate as follows:

The formula is $Al_2(SO_4)_3$

2 atoms of Al, relative atomic mass 27 $=$ 54

3 atoms of S, relative atomic mass   32 $=$ 96

12 atoms of O, relative atomic mass 16 $=$ 192

Total $= 342$

Relative formula mass of $Al_2(SO_4)_3$ $= 342$

The term relative formula mass is convenient as it can be used *for both* covalent compounds and ionic compounds. The term relative molecular mass will be used *only for* covalent compounds which consist of molecules (e.g., methane, CH_4; ethanol, C_2H_5OH). You will find both terms used in the section of examination questions.

Problems on Relative Formula Mass

Work out the relative formula masses of these compounds:

SO_2	$NaOH$	KNO_3
$MgCO_3$	$PbCl_2$	$MgCl_2$
$Mg(NO_3)_2$	$Zn(OH)_2$	$ZnSO_4$
H_2SO_4	HNO_3	$MgSO_4 \cdot 7H_2O$
$CaSO_4$	Pb_3O_4	P_2O_5
Na_2CO_3	$Ca(OH)_2$	$CuCO_3$
$CuSO_4$	$Ca(HCO_3)_2$	$CuSO_4 \cdot 5H_2O$
$Al_2(SO_4)_3$	$Na_2CO_3 \cdot 10H_2O$	$FeSO_4 \cdot 7H_2O$

Percentage Composition

From the formula of a compound, one can work out the percentage by mass of each element present in the compound.

Example 1   Calculate the percentage of silicon and oxygen in silicon(IV) oxide (silica).

Method: First, work out the relative formula mass.
The formula is SiO_2

1 atom of silicon, relative atomic mass   28 = 28

2 atoms of oxygen, relative atomic mass 16 = 32

Total = Relative formula mass   = 60

$$\text{Percentage of silicon} = \frac{28}{60} \times 100 = \frac{7}{15} \times 100$$

$$= \frac{7 \times 20}{3} = 46.7\%$$

$$\text{Percentage of oxygen} = \frac{32}{60} \times 100 = \frac{8}{15} \times 100$$

$$= \frac{8 \times 20}{3} = 53.3\%$$

Since every formula unit of silicon(IV) oxide is 46.7% silicon, and all formula units are identical, bulk samples of silicon(IV) oxide all contain 46.7% silicon. This is true whether you are talking about silicon(IV) oxide found as quartz, or amethyst or crystoballite or sand. It is an example of the *Law of Constant Composition.* This states that *all pure samples of a compound contain the same elements chemically combined in the same proportions by mass.*

In general,
Percentage of element A =

$$\frac{\text{Relative atomic mass of } A \times \text{No. of atoms of } A \text{ in formula}}{\text{Relative formula mass of compound}} \times 100$$

Example 2   Find the percentage by mass of magnesium, oxygen and sulphur in magnesium sulphate.

Method: First calculate the relative formula mass.
The formula is $MgSO_4$

1 atom of magnesium, relative atomic mass 24 =   24

1 atom of sulphur, relative atomic mass   32 =   32

4 atoms of oxygen, relative atomic mass   16 =   64

Total = Relative formula mass   = 120

$$\text{Percentage of magnesium} = \frac{\text{Rel. at. mass of Mg} \times \text{No. of Mg atoms}}{\text{Relative formula mass}} \times 100$$

$$= \frac{24}{120} \times 100 = \frac{2}{10} \times 100 = 20\%$$

$$\text{Percentage of sulphur} = \frac{\text{Rel. at. mass of S} \times \text{No. of S atoms}}{\text{Relative formula mass}} \times 100$$

$$= \frac{32}{120} \times 100 = \frac{8}{30} \times 100$$

$$= 26.7\%$$

$$\text{Percentage of oxygen} = \frac{\text{Rel. at. mass of O} \times \text{No. of O atoms}}{\text{Relative formula mass}} \times 100$$

$$= \frac{16 \times 4}{120} \times 100 = \frac{16}{30} \times 100$$

$$= 53.3\%$$

Answer: Magnesium 20%; Sulphur 26.7%; Oxygen 53.3%. You can check on the calculation by adding up the percentages to see whether they add up to 100. In this case, $20 + 26.7 + 53.3 = 100$.

Example 3 Calculate the percentage of water in copper sulphate crystals.

Method: Find the relative formula mass.
The formula is $CuSO_4 \cdot 5H_2O$
1 atom of copper, relative atomic mass $64 = 64$ (approx.)
1 atom of sulphur, relative atomic mass $32 = 32$
4 atoms of oxygen, relative atomic mass $16 = 64$
5 molecules of water, $5 \times [(2 \times 1) + 16]$ $= 5 \times 18 = 90$
Total = Relative formula mass $= 250$
Mass of water $= 90$

$$\text{Percentage of water} = \frac{\text{Mass of water in formula}}{\text{Relative formula mass}} \times 100$$

$$= \frac{90}{250} \times 100 = \frac{9}{25} \times 100$$

$$= 9 \times 4 = 36\%$$

Answer: The percentage of water in copper sulphate crystals is 36%.

Problems on Percentage Composition

Section 1

Calculators and logarithmic tables are not needed for these problems.

1. Calculate the percentages of magnesium and oxygen in magnesium oxide, using the expression

    $$\text{Percentage of element} = \frac{\text{Rel. at. mass} \times \text{No. of atoms of element}}{\text{Relative formula mass}} \times 100$$

2. Calculate the percentages by mass of calcium, carbon and oxygen in calcium carbonate.

3. Find the percentages by mass of potassium, hydrogen, carbon and oxygen in potassium hydrogencarbonate, $KHCO_3$.

4. Find the percentages by mass of
    (a) nitrogen and oxygen in nitrogen monoxide, NO
    (b) hydrogen and fluorine in hydrogen fluoride, HF
    (c) beryllium and oxygen in beryllium oxide, BeO
    (d) lithium and oxygen in lithium oxide, Li_2O.

5. Calculate the percentages by mass of
    (a) carbon and hydrogen in ethane, C_2H_6
    (b) sodium, oxygen and hydrogen in sodium hydroxide, NaOH
    (c) sulphur and oxygen in sulphur trioxide, SO_3
    (d) carbon and hydrogen in propyne, C_3H_4.

6. Calculate the percentages by mass of
    (a) carbon and hydrogen in heptane, C_7H_{16}
    (b) magnesium and nitrogen in magnesium nitride, Mg_3N_2
    (c) sodium and iodine in sodium iodide, NaI
    (d) calcium and bromine in calcium bromide, $CaBr_2$.

Section 2

These problems can be solved without the use of calculators.

1. Calculate the percentage by mass of
    (a) carbon and hydrogen in pentene, C_5H_{10}
    (b) nitrogen, hydrogen and oxygen in ammonium nitrate

    (c) iron, oxygen and hydrogen in iron(II) hydroxide

    (d) carbon, hydrogen and oxygen in ethanedioic acid, $C_2O_4H_2$.

2. Find the percentage by mass of

    (a) iron, sulphur and oxygen in iron(III) sulphate

    (b) water in chromium(III) nitrate-9-water, $Cr(NO_3)_3 \cdot 9H_2O$

    (c) water in sodium sulphide-9-water, $Na_2S \cdot 9H_2O$

    (d) silicon in silicon(IV) oxide, SiO_2.

3. Calculate the percentage of

    (a) carbon, hydrogen and oxygen in propanol, C_3H_7OH

    (b) carbon, hydrogen and oxygen in ethanoic acid, CH_3COOH

    (c) carbon, hydrogen and oxygen in methyl methanoate, $HCOOCH_3$

    (d) aluminium and sulphur in aluminium sulphide, Al_2S_3.

3. How to Calculate the Masses of Solids which will React Together

The Mole

Looking at equations tells us a great deal about chemical reactions. For example,

$$Fe(s) + S(s) \rightarrow FeS(s)$$

tells us that iron and sulphur combine to form iron(II) sulphide, and that one atom of iron combines with one atom of sulphur. Chemists are interested in the exact quantities of substances which react together in chemical reactions. For example, in the reaction between iron and sulphur, if you want to measure out just enough iron to combine with, say, 10 g of sulphur, how do you go about it? What you need to do is to count out equal numbers of atoms of iron and sulphur. This sounds a formidable task, and it puzzled a chemist called Avogadro, working in Italy early in the nineteenth century. He managed to solve this problem with a piece of clear thinking which makes the problem look very simple once you have followed his argument.

Avogadro reasoned in this way:

We know from their relative atomic masses that one atom of carbon is 12 times as heavy as an atom of hydrogen. Therefore, we can say:

If   1 atom of carbon is   12 times as heavy as 1 atom of hydrogen,
then 1 dozen C atoms are   12 times as heavy as 1 dozen H atoms,
and 1 hundred C atoms are 12 times as heavy as 1 hundred H atoms,
and 1 million C atoms are 12 times as heavy as 1 million H atoms,

and it follows that when we see a mass of carbon which is 12 times as heavy as a mass of hydrogen, the two masses must contain equal numbers of atoms. If we have 12 g of carbon and 1 g of hydrogen, we know that we have the same number of atoms of carbon and hydrogen. The same argument applies to any element. When we take the relative atomic mass of an element in grams:

40 g Calcium	24 g Magnesium	32 g Sulphur	12 g Carbon	1 g Hydrogen

all these masses contain the same number of atoms. This number is 6.023×10^{23}. The amount of an element which contains this number of atoms is called a *mole*. (The symbol for *mole* is *mol*.) The ratio 6.023×10^{23} mol^{-1} is called the *Avogadro constant*. We can count out 6×10^{23} atoms of any element by weighing out its relative atomic mass in grams. If we want to react iron and sulphur so that there is an atom of sulphur for every atom of iron, we can count out 6×10^{23} atoms of sulphur by weighing out 32 g of sulphur and we can count out 6×10^{23} atoms of iron by weighing out 56 g of iron. Since one atom of iron reacts with one atom of sulphur to form one formula unit of iron(II) sulphide, one mole of iron reacts with one mole of sulphur to form one mole of iron(II) sulphide:

$$Fe(s) + S(s) \rightarrow FeS(s)$$

and 56 g of iron react with 32 g of sulphur to form 88 g of iron(II) sulphide.

Just as a mole of an element is the relative atomic mass in grams, a mole of a compound is the relative formula mass in grams. If you want to weigh out a mole of sodium hydroxide, you first work out its relative formula mass.

The formula is NaOH

1 atom of Na, relative atomic mass $23 = 23$

1 atom of O, relative atomic mass $16 = 16$

1 atom of H, relative atomic mass $1 = 1$

Total = Relative formula mass $= 40$

If you weigh out 40 g of sodium hydroxide, you have a mole of sodium hydroxide. This is often referred to as the molar mass of the compound. The *molar mass of a compound* is the relative formula mass in grams. The *molar mass of an element* is the relative atomic mass in grams. The molar mass of sodium hydroxide is 40 g mol^{-1}; the molar mass of sodium is 23 g mol^{-1}.

Remember that most gaseous elements consist of molecules, not atoms. Chlorine exists as Cl_2 molecules, oxygen as O_2 molecules, hydrogen as H_2 molecules, and so on. To work out the mass of a mole of chlorine, you must use the relative formula mass (or relative molecular mass) of Cl_2.

Relative atomic mass of chlorine $= 35.5$

Relative formula mass $\qquad = 2 \times 35.5 = 71$

Mass of 1 mole of chlorine $\quad = 71$ grams.

The noble gases, helium, neon, argon, krypton and xenon, exist as atoms. Since the relative atomic mass of helium is 4, the mass of 1 mole of helium is 4 g.

Problems on the Mole

REMEMBER:

$$\text{Number of moles of an element} = \frac{\text{Mass of the element}}{\text{Relative atomic mass of the element}}$$

$$\text{Number of moles of a compound} = \frac{\text{Mass of the compound}}{\text{Relative formula mass of the compound}}$$

Section 1

1. What are the relative atomic masses of sodium, magnesium and lead? What is the mass of 1 mole of:

   (a) sodium   (b) magnesium   (c) lead?

2. What are the relative atomic masses of barium, chromium and tin? What is the mass of:

   (a) 0.1 mole of barium   (b) 0.1 mole of chromium
   (c) 0.1 mole of tin?

3. Use the relative atomic masses of the elements to calculate the mass of:

   (a) 2 moles of iodine molecules (b) 2 moles of silver
   (c) 2 moles of aluminium   (d) 2 moles of mercury.

4. Calculate the mass of 0.25 mole of each of these elements:

   (a) silver   (b) sulphur   (c) magnesium
   (d) calcium   (e) neon.

5. Use the relative atomic masses to find the number of moles of the element in:
   (a) 54 g aluminium   (b) 1.6 g sulphur   (c) 42 g iron
   (d) 54 g silver   (e) 13 g zinc.

6. What are the relative formula masses of the following compounds: carbon dioxide, sulphuric acid, hydrogen chloride and sodium hydroxide? State the mass of:
   (a) 1 mole of carbon dioxide   (b) 1 mole of sulphuric acid
   (c) 1 mole of hydrogen chloride   (d) 1 mole of sodium hydroxide.

7. Use the relative formula masses of the compounds to calculate the mass of:
   (a) 1 mole of sodium chloride   (b) 0.5 mole of potassium hydroxide
   (c) 4 moles of iron(II) chloride   (d) 2.5 moles of sodium carbonate
   (e) 0.1 mole of zinc chloride.

8. For each of the following compounds, work out the relative formula mass, and then state (a) the mass of one mole of the compound, and (b) the mass of 0.25 mole of the compound:
   calcium chloride, copper carbonate, barium hydroxide, sodium nitrate.

Section 2

1. State the mass of each element in:
   (a) 0.5 mole chromium   (b) 1/7 mole iron
   (c) 1/3 mole carbon   (d) 1/4 mole magnesium
   (e) 1/7 mole nitrogen molecules   (f) 1/4 mole oxygen molecules
   Remember that nitrogen and oxygen exist as diatomic molecules, N_2 and O_2.

2. Calculate the number of moles of each element in:
   (a) 46 g sodium   (b) 130 g zinc   (c) 10 g calcium
   (d) 2.4 g magnesium   (e) 8 g sulphur.

3. Calculate the mass of:
   (a) 1 mole of sodium atoms   (b) ½ mole of nitrogen atoms
   (c) ½ mole of nitrogen molecules   (d) ¼ mole of sulphur atoms
   (e) 0.2 mole of bromine atoms   (f) 0.2 mole of bromine molecules.

4. Calculate the number of moles of atoms in:

    (a) 23 g sodium (b) 64 g sulphur (c) 9 g aluminium

    (d) 120 g calcium (e) 12 g magnesium (f) 7 g iron.

5. Find the mass of each element in:

    (a) 10 moles of lead (b) 1/6 mole copper

    (c) 0.1 mole iodine molecules (d) 10 moles hydrogen molecules

    (e) 0.25 mole calcium (f) 0.25 mole bromine molecules

    (g) ¾ mole iron (h) 0.20 mole zinc

    (i) ½ mole chlorine molecules (j) 0.1 mole neon.

6. State the number of moles in:

    (a) 58.5 g sodium chloride (b) 26.5 g anhydrous sodium carbonate

    (c) 50.0 g calcium carbonate (d) 15.9 g copper(II) oxide

    (e) 8.0 g sodium hydroxide (f) 303 g potassium nitrate

    (g) 9.8 g sulphuric acid (h) 499 g copper sulphate-5-water.

7. Given Avogadro's constant is 6×10^{23} mol^{-1}, calculate the number of atoms in:

    (a) 35.5 g of chlorine (b) 27g of aluminium

    (c) 3.1 g of phosphorus (d) 336 g of iron

    (e) 48 g of magnesium (f) 1.6 g of oxygen

    (g) 0.4 g of oxygen (h) 216 g of silver.

8. How many grams of zinc contain:

    (a) 6×10^{23} atoms (b) 6×10^{20} atoms?

9. How many grams of aluminium contain:

    (a) 2×10^{23} atoms (b) 6×10^{20} atoms?

10. What mass of carbon contains:

    (a) 6×10^{23} atoms (b) 2×10^{21} atoms?

11. Write down:

    (a) the mass of calcium which has the same number of atoms as 12 g of magnesium

    (b) the mass of silver which has the same number of atoms as 3 g of aluminium

    (c) the mass of zinc with the same number of atoms as 1 g of helium

    (d) the mass of sodium which has 5 times the number of atoms in 39 g of potassium.

Calculation of Mass of Reactant or Mass of Product

We can use the idea of the mole to work out what mass of product we shall obtain by starting with a known mass of reactant.

Example 1   What mass of magnesium oxide is obtained from the complete combustion of 12 g of magnesium?

Method: First, we write the equation for the reaction:

$$2Mg(s) + O_2(g) \rightarrow 2MgO(s)$$

The equation tells us that:

2 atoms of magnesium form 2 'formula units' of magnesium oxide

therefore 1 atom of magnesium forms 1 'formula unit' of magnesium oxide

and 1 mole of magnesium atoms forms 1 mole of magnesium oxide 'formula units'.

We can therefore write

RAM in grams of magnesium forms RFM in grams of magnesium oxide

(RAM = relative atomic mass; RFM = relative formula mass)

RAM of Mg = 24

RFM of MgO = (RAM of Mg + RAM of O) = (24 + 16) = 40

Therefore, 24 g of magnesium form 40 g of magnesium oxide.

Therefore, 12 g of magnesium form 20 g of magnesium oxide.

Answer: 20 g of magnesium oxide are formed by the complete combustion of 12 g of magnesium.

Example 2   What mass of zinc sulphate can be obtained from the reaction of 10 g of zinc with an excess of dilute sulphuric acid?

Method: First, we write the equation:

$$Zn(s) + H_2SO_4(aq) \rightarrow H_2(g) + ZnSO_4(aq)$$

Since   1 mole of Zn atoms + 1 mole of H_2SO_4 'formula units' → 1 mole of H_2 molecules + 1 mole of $ZnSO_4$ 'formula units'

We can write, RAM in g of zinc forms RFM in g of zinc sulphate.

(We will leave out the information about sulphuric acid and hydrogen because this question does not ask for it.)

Calculate the relative formula mass of zinc sulphate:

1 atom of zinc (RAM 65) $\quad = \quad 65$

1 atom of sulphur (RAM 32) $= \quad 32$

4 atoms of oxygen (RAM 16) $= \quad 64$

Total = RFM of $ZnSO_4$ $\quad = \quad 161$

Since the RAM in g of zinc forms the RFM in g of zinc sulphate,

then $\qquad$ 65 g of zinc forms 161 g of zinc sulphate

therefore $\qquad$ 1 g of zinc forms $\dfrac{161}{65}$ g of zinc sulphate

and $\qquad$ 10 g of zinc forms $\dfrac{161}{65} \times 10$ g of zinc sulphate

$$= 24.8 \text{ g of zinc sulphate}$$

Answer: 24.8 g of zinc sulphate can be obtained from 10 g of zinc.

The calculation in this problem is a ratio type of calculation. It is tackled in the same way as you tackle such problems in your mathematics lessons, by the unitary method. For example, take the question:

If 3 packets of crisps cost 24 p, what is the cost of 7 packets of crisps?
You work out: If 3 packets of crisps cost 24 p

1 packet of crisps costs $\dfrac{24}{3}$ p

and 7 packets of crisps cost $\dfrac{24}{3} \times 7 = 56$ p

Example 3 Calculate the mass of carbon dioxide produced by heating 15 g of limestone.

Method: The equation

$$CaCO_3(s) \rightarrow CaO(s) + CO_2(g)$$

tells us that 1 mole of calcium carbonate forms 1 mole of carbon dioxide. Then RFM in g of calcium carbonate forms RFM in g of carbon dioxide.

Using the RAM Ca = 40, C = 12, O = 16,

RFM of calcium carbonate = $[40+12+(3\times16)]$ = 100

RFM of carbon dioxide   = $[12+(2\times16)]$ = 44

Since RFM in g of calcium carbonate forms RFM in g of carbon dioxide,

100 g of calcium carbonate form 44 g of carbon dioxide

1 g of calcium carbonate forms $\dfrac{44}{100}$ g of carbon dioxide

15 g of calcium carbonate form $\dfrac{44}{100}\times15$ g of carbon dioxide

$$= 6.6 \text{ g of carbon dioxide}$$

Answer: 6.6 g of carbon dioxide will be produced.

Example 4 If 4.2 g of sodium hydrogencarbonate are heated, what mass of anhydrous sodium carbonate will be formed?

Method: First, write the equation:

$$2NaHCO_3(s) \rightarrow Na_2CO_3(s) + CO_2(g) + H_2O(g)$$

This shows that 2 moles of sodium hydrogencarbonate give 1 mole of sodium carbonate. Therefore $2\times$RFM in g of sodium hydrogencarbonate give RFM in g of sodium carbonate.

RFM of $NaHCO_3$ = $23+1+12+(3\times16)$ = 84

RFM of Na_2CO_3 = $(2\times23)+12+(3\times16)$ = 106

2×84 g sodium hydrogencarbonate give 106 g sodium carbonate.

If 168 g sodium hydrogencarbonate give 106 g sodium carbonate

1 g sodium hydrogencarbonate gives $\dfrac{106}{168}$ g sodium carbonate

4.2 g sodium hydrogencarbonate give $\dfrac{4.2\times106}{168}$ g sodium carbonate

$$= \frac{0.6\times106}{24} = \frac{0.1\times106}{4}$$

$$= 2.65 \text{ g}$$

Answer: 2.65 g of sodium carbonate will be formed.

Using the Masses of the Reactants to work out the Equation for a Reaction

The equation for a reaction can be used to enable you to calculate the masses of chemicals taking part in the reaction. The converse is also true. If you know the mass of each substance taking part in a reaction, you can calculate the number of moles of each substance taking part in the reaction, and this will tell you the equation.

Example 1   Iron burns in chlorine to form iron chloride. An experiment showed that 5.6 g of iron combined with 10.65 g of chlorine. Deduce the equation for the reaction.

Method:
5.6 g of iron combine with 10.65 g of chlorine

Relative atomic masses are Fe = 56;   Cl = 35.5

Number of moles of iron   = 5.6/56   = 0.1

Number of moles of chlorine   = 10.65/35.5 = 0.3

The equation must be:

$$Fe + 3Cl \rightarrow$$

Since chlorine exists as Cl_2 molecules, we must multiply by 2:

$$2Fe + 3Cl_2 \rightarrow$$

To balance the equation, the right-hand side must read $2FeCl_3$.

Therefore,

Answer: $2Fe(s) + 3Cl_2(g) \rightarrow 2FeCl_3(s)$

Example 2   17 g of sodium nitrate react with 19.6 g of sulphuric acid to give 12.6 g of nitric acid. Deduce the equation for the reaction.

Method:
Relative formula masses are: $NaNO_3 = 85$, $H_2SO_4 = 98$, $HNO_3 = 63$

Number of moles of $NaNO_3$ = 17/85   = 0.2

Number of moles H_2SO_4 = 19.6/98 = 0.2

Number of moles of HNO_3 = 12.6/63 = 0.2

0.2 mole $NaNO_3$ reacts with 0.2 mol H_2SO_4 to form 0.2 mol of HNO_3

1   mole $NaNO_3$ reacts with 1   mol H_2SO_4 to form 1   mol of HNO_3

8. What mass of zinc chloride is formed when 13 g zinc are completely converted to chloride?

9. Calculate the mass of potassium chloride formed when a solution containing 8 g potassium hydroxide is neutralised with hydrochloric acid.

$$KOH(aq) + HCl(aq) \rightarrow KCl(aq) + H_2O(l)$$

10. Calculate how much sodium nitrate you need to give 126 g of nitric acid by the reaction

$$NaNO_3(s) + H_2SO_4(l) \rightarrow HNO_3(l) + NaHSO_4(s)$$

11. Which one of the following contains the same number of atoms as 7 g of iron?

    (a) 4 g of aluminium   (b) 4 g of magnesium   (c) 4 g of sulphur
    (d) 3 g of carbon   (e) 4 g of calcium

12. State which of these compounds contains the largest percentage by mass of nitrogen:

    (a) ammonium chloride, NH_4Cl (b) ammonium nitrate, NH_4NO_3

    (c) ammonium sulphate, $(NH_4)_2SO_4$ (d) ammonia, NH_3
    (e) urea, $CO(NH_2)_2$

13. Which of the following contains the same number of atoms as 10 g of calcium?

    (a) 6 g of sodium   (b) 13 g of chromium   (c) 8 g of magnesium
    (d) 26 g of silver (e) 7 g of aluminium

Section 3   Problems from GCE O-level Papers

1. (a) Describe how you would prepare dry crystals of hydrated copper(II) sulphate ($CuSO_4 \cdot 5H_2O$) using copper(II) oxide and dilute sulphuric acid as the starting materials.

    Calculate the mass (in grams) of the hydrated crystals which could be obtained from 0.01 mole of copper(II) oxide.

   (b) Describe what happens when the hydrated crystals are gently heated until no further change in mass occurs.

    Calculate the percentage change in mass of the hydrated crystals.
    (O & C)

2. (a) Describe how you would prepare dry crystals of hydrated sodium sulphate ($Na_2SO_4 \cdot 10H_2O$), starting from dilute aqueous sodium hydroxide and dilute sulphuric acid.

(b) Calculate the mass, in grams, of the hydrated crystals which you would expect to obtain from 0.05 mole of sodium hydroxide.

(O & C)

3. (a) Draw a labelled diagram of an apparatus which could be used to find the formula of water by reducing dry copper(II) oxide with dry hydrogen.

(b) In such an experiment the following results were obtained:

Mass of copper(II) oxide before experiment = 15.80 g

Mass of copper after experiment   = 12.60 g

Mass of water collector before experiment = 65.06 g

Mass of water collector after experiment   = 68.66 g

Use these results to determine the formula of water.

(c) Referring to your calculation, suggest why it was deduced from earlier work involving this experiment that the relative atomic masses of hydrogen and oxygen were 1 and 8 respectively.

(O & C)

4. (a) Describe how you would prepare dry crystals of hydrated zinc sulphate ($ZnSO_4 \cdot 7H_2O$), starting from zinc oxide and dilute sulphuric acid.

(b) Calculate the mass, in grams, of the *hydrated* crystals which could, theoretically, be obtained from 0.05 mole of zinc oxide.

(c) 0.01 mole of the zinc sulphate crystals was dissolved in water and to this solution excess aqueous barium chloride was added. Describe what would be seen, write an ionic equation for the reaction and calculate the mass, in grams, of the dry solid product which could be obtained. (O & C)

5. Calculate the percentage of nitrogen in pure ammonium nitrate, NH_4NO_3. (O & C)

6. Copper(II) sulphate crystals have the formula $CuSO_4 \cdot 5H_2O$. What mass of crystals could theoretically be obtained from 8.0g of copper(II) oxide? (O & C)

7. Describe how you would make reasonably dry crystals of hydrated zinc sulphate ($ZnSO_4 \cdot 7H_2O$), starting from zinc oxide.

Calculate the theoretical yield of the hydrated crystals from 4.05 g of the oxide. (O & C)

8. Copper(II) sulphate crystals, $CuSO_4 \cdot 5H_2O$, are prepared from copper(II) oxide. What mass of the oxide would you theoretically require in order to produce 1000 g of the crystals? ($Cu = 64$)

   (O & C)

9. A recently developed process for reducing the amount of nitrogen oxide in vehicle exhaust gases involves the injection of a stream of ammonia gas into the combustion products. The reaction is believed to be

$$4NH_3 + 6NO \rightarrow 5N_2 + 6H_2O$$

   Assuming that the average small car emission is 1.8 g of NO per mile, calculate the mass of ammonia which would be needed to react with the NO emitted by a car travelling 10 000 miles in one year.

   (Oxford)

10. 3.22 g of an efflorescent salt hydrate of formula $M_2SO_4 \cdot 10H_2O$ was placed in a dry atmosphere until no further loss in mass occurred; the mass of the residue was 1.42 g. Calculate the relative atomic mass of M.

   (Oxford)

11. (a) A power station which it is proposed to build in the USA would burn about 50 000 tons of coal in a single day. Assuming this figure is equivalent to about 48 000 tons of fossil carbon consumed in the reaction $C + O_2 \rightarrow CO_2$, calculate the daily tonnage of carbon dioxide which would be added to the atmosphere by this plant alone.

    (b) A motorist drives 10 miles to work and 10 miles back on 250 working days in a year. His car has a fuel consumption of 20 miles to the gallon. How much lead does the motorist discharge into the environment, assuming the lead content of petrol is 2 g/gallon?

    (Oxford)

12. If ammonium sulphate, $(NH_4)_2SO_4$, and sodium nitrate, $NaNO_3$, cost the same per ton, which is the cheaper source of nitrogen?

    (London)

13. Sodium hydrogencarbonate decomposes on heating according to the equation

$$2NaHCO_3 \rightarrow Na_2CO_3 + H_2O + CO_2$$

    A loss of 6.2 g was observed when 25.0 g of a mixture of sodium carbonate and sodium hydrogencarbonate were heated to constant weight. Calculate the mass of each salt in the original mixture.

    (Oxford)

14. The relative atomic masses of zinc and iodine are 65 and 127 respectively. From this statement, it may be deduced that

(a) the radii of the atoms of zinc and iodine are in the ratio of 65:127

(b) 65 g of zinc occupies the same volume as 127 g of iodine

(c) 65 atoms of zinc weigh the same as 127 atoms of iodine

(d) 1 g of zinc contains 65 atoms and 1 g of iodine contains 127 atoms

(e) 65 g of zinc contains the same number of atoms as 127 g of iodine.

Choose the correct answer.   (London)

15. The ratio of the numbers of atoms present in 20 g samples of neon (relative atomic mass = 20), argon (relative atomic mass = 40), and bromine (relative atomic mass = 80) would be respectively

(a) 1:1:1   (b) 1:2:2   (c) 1:2:4

(d) 4:2:1   (e) 8:4:1

Choose the correct answer.   (London)

16. To obtain potassium hydrogensulphate from sulphuric acid and potassium hydroxide, it is necessary to have

(a) equal masses of acid and alkali

(b) 1 mol of acid to 1 mol of alkali

(c) 1 mol of acid to 2 mol of alkali

(d) 2 mol of acid to 1 mol of alkali

(e) a large excess of acid.   (Cambridge)

17. One mole of magnesium was heated in a closed vessel in the presence of one mole of chlorine (Cl_2) and one mole of bromine (Br_2). If the solid remaining was soluble in water and had a mass of 139.5 g, it could be deduced that

(a) all of the chlorine had reacted

(b) all of the bromine had reacted

(c) some magnesium was left

(d) the mass of halogen left was 115.5 g

(e) the solid consisted only of the compound MgBrCl.   (Cambridge)

18. One mole of carbon-12 has a mass of

(a) 0.012 kg   (b) 0.0224 kg

(c) 0.024 kg   (d) 1 kg

(e) 12 kg   (Cambridge)

19. Four separate solutions are prepared so that each solution contains 1 g of one of the ions Fe^{2+}, Fe^{3+}, Cu^{2+}, Zn^{2+}. To each solution is added an excess of sodium hydroxide solution and the mass of any resulting precipitate is found. Which one of the following diagrams (a), (b), (c), (d) or (e) illustrates the results?

[Relative atomic masses: H, 1; O, 16; Fe, 56; Cu, 64; Zn, 65]

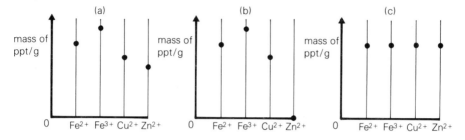

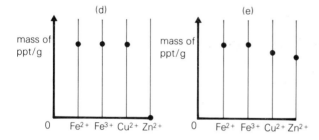

(Cambridge)

4. Empirical Formulae and Molecular Formulae

Empirical Formulae

From the formula, one can find the mass of each element present in a certain mass of the compound. The reverse is also true: from the mass of each element present in a sample of the compound, one can find the formula of the compound. The method uses the ratio type of calculation. For example:

If the relative atomic mass of magnesium is 24, how many moles of magnesium are there in 6 g of magnesium?

If 24 g of magnesium are 1 mole,
then 6 g of magnesium are 6/24 mole = ¼ mole.

In general,

$$\text{No. of moles of element} = \frac{\text{Mass of element}}{\text{Relative atomic mass}}$$

$$\text{No. of moles of compound} = \frac{\text{Mass of compound}}{\text{Relative formula mass of compound}}$$

Example 1 Given that 127 g of copper combine with 32 g of oxygen, what is the formula of copper oxide?

Elements	Copper		Oxygen
Symbols	Cu		O
Masses	127 g		32 g
Relative atomic masses	63.5		16
Number of moles	$\dfrac{127}{63.5}$		$\dfrac{32}{16}$
	= 2		= 2
Divide through by 2	= 1 mole	to	1 mole
Number of atoms	= 1 atom	to	1 atom
Formula		CuO	

We divide through by two to obtain the simplest formula for copper oxide which will fit the data. *The simplest formula which represents the composition of a compound is called the empirical formula.*

Example 2 Given that 0.96 g of magnesium combines with 2.84 g of chlorine, what is the empirical formula for magnesium chloride?

Elements	*Magnesium*		*Chlorine*
Symbols	Mg		Cl
Masses	0.96 g		2.84 g
Relative atomic masses	24		35.5
Number of moles	$\dfrac{0.96}{24}$		$\dfrac{2.84}{35.5}$
	= 0.04		= 0.08
Divide through by the smaller number	= 1 mole	to	2 moles
Number of atoms	= 1 atom	to	2 atoms
Empirical formula		$MgCl_2$	

Example 3 If the percentage of water in magnesium sulphate crystals is 51.2%, what is n in the formula $MgSO_4 \cdot nH_2O$?

Note that, when we say that the percentage of water in the crystals is 51.2%, we mean that 100 g of crystals contain 51.2 g of water. The difference, 48.8 g is the mass of magnesium sulphate.

Compounds	*Magnesium sulphate*		*Water*
Formulae	$MgSO_4$		H_2O
Masses	48.8 g		51.2 g
Relative formula masses	120		18
Number of moles	$\dfrac{48.8}{120}$		$\dfrac{51.2}{18}$
	= 0.406		= 2.85
Divide through by the smaller number	= 1 mole	to	7 moles
Empirical formula		$MgSO_4 \cdot 7H_2O$	

Example 4 When 127 g of copper combine with oxygen, 143 g of an oxide are formed. What is the empirical formula of the oxide?

Method: You will notice here that the mass of oxygen is not given to you. It is obtained by subtraction.

Mass of copper $= 127$ g

Mass of oxide $= 143$ g

Mass of oxygen $= 143 - 128 = 16$ g.

Now you can carry on as before:

Elements	Copper	Oxygen
Symbols	Cu	O
Masses	127 g	16 g
Relative atomic masses	63.5	16
Number of moles	$\dfrac{127}{63.5}$	$\dfrac{16}{16}$
	$= 2$	$= 1$
Number of atoms	$= 2$ to	1

Answer: Empirical formula Cu_2O

How to find the Molecular Formula from the Empirical Formula

The molecular formula of a compound can be found from the empirical formula if the relative molecular mass (or relative formula mass) is known.

Example 1 Analysis shows the empirical formula of a compound to be CH_2O. Its relative molecular mass is 60. What is its molecular formula?

Method:

Relative molecular mass $= 60$

Relative empirical formula mass $= (12 + 2 + 16) = 30$

The relative molecular mass is double the relative empirical formula mass.

The molecular formula is double the empirical formula.

Answer: The molecular formula is $C_2H_4O_2$.

Example 2 What is the molecular formula of the compound, A, which has an empirical formula C_2H_6O and a relative molecular mass of 46?

Method:

Relative molecular mass of A   = 46

Relative empirical formula mass = $(2 \times 12) + 6 + 16 = 46$

Answer: The empirical formula gives the correct relative molecular mass; therefore the molecular formula is the same as the empirical formula, C_2H_6O.

xample 3 The empirical formula of a liquid, B, is C_2H_4O. The relative molecular mass is 88. What is the molecular formula of B?

Method:

Relative empirical formula mass of B = $(2 \times 12) + 4 + 16 = 44$

Since relative molecular mass $= 88 = 2 \times$ relative empirical formula mass,

molecular formula $= 2 \times$ empirical formula

                                $= C_4H_8O_2$

Answer: The molecular formula is $C_4H_8O_2$.

roblems on Empirical Formulae and Molecular Formulae

ection 1

These examples can be done without calculators or logarithms.

1. 2.3 g of sodium combine with 0.80 g of oxygen.
   How many moles of sodium does this mass represent?
   How many moles of oxygen are involved?
   How many moles of sodium combine with one mole of oxygen?
   What is the formula of sodium oxide?

2. 0.72 g of magnesium combine with 0.28 g of nitrogen.
   How many moles of magnesium does this represent?
   How many moles of nitrogen combine?
   How many moles of magnesium combine with one mole of nitrogen atoms?
   What is the formula of magnesium nitride?

3. 1.68 g of iron combine with 0.64 g of oxygen.
   How many moles of iron does this mass represent?
   How many moles of oxygen combine?

How many moles of iron combine with one mole of oxygen atoms? What is the formula of this oxide of iron?

4. A compound contains 55.5% by mass of mercury; 44.5% by mass of bromine.
   How many grams of mercury are there in 100 g of the compound?
   How many grams of bromine are there in 100 g of the compound?
   How many moles of mercury are there in 100 g of the compound?
   How many moles of bromine are there in 100 g of the compound?
   Work out the ratio of moles of bromine to moles of mercury.
   What is the formula of the compound?

5. Calculate the empirical formula of the compound formed when 2.7 g of aluminium form 5.10 g of its oxide.
   What is the mass of aluminium?
   What is the mass of oxygen (not oxide)?
   How many moles of aluminium combine?
   How many moles of oxygen combine?
   What is the ratio of moles of aluminium to moles of oxygen?
   What is the formula of aluminium oxide?

6. Barium chloride forms a hydrate which contains 85.25% barium chloride and 14.75% water of crystallisation. What is the formula of this hydrate?
   What is the mass of barium chloride in 100 g of the hydrate?
   What is the mass of water in 100 g of the hydrate?
   What is the relative formula mass of barium chloride?
   What is the relative formula mass of water?
   How many moles of barium chloride are present in 100 g of the hydrate?
   How many moles of water are present in 100 g of the hydrate?
   What is the ratio of moles of barium chloride to moles of water?
   What is the formula of barium chloride hydrate?

7. Calculate the empirical formula of the compound formed when 414 g of lead form 478 g of a lead oxide.
   What mass of lead is present?
   How many moles of lead are present?
   What mass of oxygen (not oxide) is present?
   How many moles of oxygen are present?
   What is the formula of this oxide of lead?

8. Calculate the empirical formulae of the compounds which analyse as
   (a) 84% carbon, 16% hydrogen
   (b) 72% magnesium, 28% nitrogen
   (c) 36% aluminium, 64% sulphur

    (d) 20% calcium, 80% bromine

    (e) 52% chromium, 48% sulphur

Section 2

No calculators or logarithms are required.

1. Calculate the empirical formulae of the compounds containing
   (a) sulphur 50%, oxygen 50%
   (b) sulphur 40%, oxygen 60%
   (c) nitrogen 47%, oxygen 53%
   (d) nitrogen 30.5%, oxygen 69.5%
   (e) carbon 75%, hydrogen 25%
   (f) carbon 85.7%, hydrogen 14.3%

2. Calculate the empirical formulae of the following compounds:
   (a) 0.62 g of phosphorus combined with 0.48 g of oxygen
   (b) 1.4 g of nitrogen combined with 0.30 g of hydrogen
   (c) 0.62 g of lead combined with 0.064 g of oxygen
   (d) 3.5 g of silicon combined with 4.0 g of oxygen
   (e) 1.10 g of manganese combined with 0.64 g of oxygen
   (f) 4.2 g of nitrogen combined with 12.0 g of oxygen
   (g) 2.6 g of chromium combined with 5.3 g of chlorine

3. Find the molecular formula for each of the following compounds from the empirical formula and the relative molecular (or formula) masses:

	Empirical formula	RMM		Empirical formula	RMM
A	C_2H_6O	46	E	CH_2	42
B	C_2H_4O	88	F	CH_3O	62
C	CH_3	30	G	CH_2Cl	99
D	CH	78	H	C_2HNO_2	213

4. Calculate the empirical formulae of the compounds formed when
   (a) 0.69 g of sodium forms 0.93 g of an oxide of sodium
   (b) 10.35 g of lead form 11.41 g of an oxide of lead

(c) 0.035 g of nitrogen forms 0.115 g of an oxide of nitrogen

(d) 2.54 g of copper form 2.86 g of an oxide

(e) 11.2 g of iron form 25.4 g of a chloride of iron

(f) 14.0 g of iron combine with 26.6 g of chlorine

5. Calculate the empirical formulae of the compounds formed when

(a) 0.24 g of carbon combines with 0.64 g of oxygen

(b) 20.7 g of lead form 23.9 g of a lead oxide

(c) 15.9 g of copper combine with 17.7 g of chlorine

(d) 6 g of magnesium combine with 4 g of oxygen

(e) 1.8 g of magnesium form 2.5 g of magnesium nitride

(f) 9 g of aluminium form 89 g of aluminium bromide.

6. Calculate the empirical formulae of these hydrates:

(a) magnesium sulphate crystals, which contain 48.8% of magnesium sulphate and 51.2% of water

(b) copper sulphate crystals, which contain 63.9% of copper sulphate and 36.1% of water

(c) crystals of chromium(III) nitrate, which contain 59.5% of chromium(III) nitrate and 40.5% of water.

7. Calculate the empirical formulae of the compounds with the following compositions:

(a) 20% magnesium, 26.6% sulphur, 53.3% oxygen

(b) 35% nitrogen, 5% hydrogen, 60% oxygen

(c) 60% carbon, 13.3% hydrogen, 26.7% oxygen

(d) 40% carbon, 6.7% hydrogen, 53.3% oxygen

Section 3 Problems from GCE O-level Papers

1. A hydrocarbon contains 80% of carbon, and its relative molecular mass is 30. Calculate its empirical formula. Write down (a) its molecular and (b) structural formulae. (O & C)

2. 1.23 g of hydrated magnesium sulphate were placed in a crucible and heated to constant mass. The loss in mass was 0.63 g. Calculate the empirical formula of the crystals. (O & C)

3. The empirical formula of butane is C_2H_5, and its relative molecular mass is 58. What is its molecular formula? Write down one possible structural formula.

For benzene, the empirical formula is CH, and the relative molecular mass is 78. Write down the molecular formula and a possible structural formula. (O & C)

4. On analysis, a hydrocarbon was found to contain 85.7% of carbon. Calculate its empirical formula. If the relative molecular mass of the hydrocarbon is 28, (a) what is its molecular formula? (b) give the structural formula of this hydrocarbon. (O & C)

5. Two oxides of a metal M (relative atomic mass 207) contain 7.18% and 13.4% of oxygen respectively. Calculate the empirical formulae of the two oxides. (O & C)

6. 0.61 g of barium chloride crystals, $BaCl_2 \cdot xH_2O$, are heated to constant mass. Water of crystallisation is given off, and the mass of the anhydrous chloride is 0.52 g. Calculate the empirical formula of the crystals. (O & C)

7. A hydrocarbon contains 7.7% of hydrogen. Calculate its empirical formula. Give a possible structural formula for this hydrocarbon. (O & C)

8. 4.17 g of a chloride of phosphorus was found to contain 3.55 g of chlorine. Find the simplest (empirical) formula for the chloride. $(A_r(P) = 31; A_r(Cl) = 35.5.)$ (Welsh JEC)

9. (a) 5 g of an oxide (A) of chromium were found on reduction to give 2.6 g of chromium. Find the simplest (or empirical) formula of the oxide. $(A_r(Cr) = 52; A_r(O) = 16.)$

   (b) Oxide (A) on strong heating forms oxygen and a second oxide of chromium (B), which is found to contain 24 g of oxygen combined with the molar mass of chromium. Find the simplest formula of oxide (B).

   (c) Using the simplest formulae for the two oxides, write an equation for the formation of oxide (B) from oxide (A). (Welsh JEC)

10. An organic compound (relative molecular mass = 28) contains 85.7% carbon and 14.3% hydrogen by mass. What are:
    (a) its empirical formula   (b) its molecular formula
    (c) its structural formula   (d) its name
    (e) the name of the series of which it is a member? (AEB 1978)

11. (a) An oxide (A) of the element molybdenum (Mo) is obtained by heating the sulphide MoS_2 in air. 7.2 g of oxide (A) is found by analysis to contain 2.4 g of oxygen.

    (i) Calculate the simplest (or empirical) formula of oxide (A) $(A_r(Mo) = 96; A_r(O) = 16.)$

(ii) Using the simplest formula for (A), write an equation for its formation from the sulphide MoS_2, given that the only other product is sulphur dioxide.

(b) When the oxide (A) is heated under special conditions with molybdenum, a second oxide (B) is formed. Oxide (B) is found to contain 32 g of oxygen combined with the molar mass of molybdenum.

(i) Calculate the simplest formula for oxide (B).

(ii) Write the equation for the formation of oxide (B) from the reaction of molybdenum with oxide (A).   (Welsh JEC)

12. In carrying out an experiment with hydrated magnesium sulphate, $MgSO_4 \cdot xH_2O$, the following results were obtained:

(1) Mass of crucible plus lid   = 14.636 g

(2) Mass of crucible plus lid plus hydrated magnesium sulphate
                                                                                    = 15.374 g

(3) Mass of crucible plus lid plus crystals after heating   = 14.996 g

(4) Mass of crucible plus lid plus crystals after further heating
                                                                                    = 14.996 g

(a) What was the mass of water of crystallisation in the sample of hydrated crystals used?

(b) What was the mass of anhydrous magnesium sulphate formed?

(c) What is x in the formula $MgSO_4 \cdot xH_2O$?

(d) Why was step (4) necessary?   (Oxford)

13. A polymer is shown by analysis to have the following composition: C, 85.7%; H, 14.3%. Calculate the empirical formula for this substance.   (Oxford)

14. 2.7 g of aluminium produced 13.35 g of aluminium chloride. What is the simplest formula of the salt?   (Oxford)

15. Two saturated hydrocarbons have the same relative molecular mass, 58. Each contains 82.8% carbon.

For each of the two hydrocarbons, write down (a) its molecular formula, (b) its structural formula.   (Oxford)

16. The element titanium combines with oxygen in two different ways. Complete the table below. (Relative atomic masses: Ti = 48.0, O = 16.)

	Oxide 1	Oxide 2
Mass of oxygen combining with 1 g of titanium	0.50 g	0.67 g
Mass of oxygen combining with 1 mole of titanium atoms		
Number of moles of oxygen atoms combining with 1 mole of titanium atoms		
Formula of oxide		

(London)

17. 0.016 g of a hydrocarbon Z was *completely* oxidised by heating with copper(II) oxide. The water formed was absorbed by anhydrous calcium chloride, and the carbon dioxide formed was collected. After cooling to room temperature, the carbon dioxide was measured as 30 cm^3.

(a) (i)   How many moles of carbon dioxide were formed?
    (ii) How many moles of carbon atoms does this contain?
    (iii) What mass of carbon does this represent?
    (iv) What mass of hydrogen was combined with this mass of carbon?
    (v)   How many moles of hydrogen atoms does this represent?

(b) In another such experiment, it was found that the ratio of carbon atoms to hydrogen atoms was 0.007 to 0.005.
    (i)   What would be the simplest formula of the hydrocarbon?
    (ii) Given that the hydrocarbon has a relative molecular mass of 178, what is its molecular formula?   (London)

18. What is (a) the empirical formula, (b) the molecular formula of a compound containing 4.04% hydrogen, 24.24% carbon and 71.72% chlorine, and relative molar mass of 99?   (SUJB)

19. The simplest formula of a hydrocarbon is CH_2 and its relative molecular mass is 112. (H = 1, C = 12). Its molecular formula is

(a) C_2H_4 (b) C_4H_8 (c) C_6H_{12} (d) C_8H_{16} (e) $C_{10}H_{20}$
(London)

20. When 5.40 g of an oxide of uranium (U) were reduced, 4.76 g of uranium metal were obtained. Calculate the simplest formula for the oxide. (Relative atomic masses: O = 16; U = 238.)   (Cambridge)

21. The mass of one mole of a chloride formed by a metal Y is 74.5 g. The formula of the chloride could be

(a) Y_3Cl (b) Y_2Cl (c) YCl (d) YCl_2 (e) YCl_3

(Cambridge)

22. A compound containing only the elements carbon and hydrogen has 80.0% by mass of carbon. Its empirical formula is

(a) C_4H (b) C_3H (c) CH_3 (d) CH_4 (e) C_2H_6

(Cambridge)

23. A compound contains the following percentage composition by mass: calcium, 29.4%; sulphur, 23.5%; oxygen, 47.1%. Calculate the simplest formula for this compound.

(Cambridge)

24. A compound is found to have the composition by mass: silicon, 87.5%; hydrogen, 12.5%. The empirical formula of the compound is

(a) SiH_2 (b) SiH_3 (c) SiH_4 (d) Si_2H_6 (e) Si_3H_8

(Cambridge)

5. Calculations on Reacting Volumes of Gases

Calculating the Reacting Volumes of Gases from the Equation for a Reaction

An intriguing feature of reactions between gases was noticed by a French chemist called *Gay Lussac* in 1808. *Gay Lussac's Law states that when gases combine they do so in volumes which bear a simple ratio to one another and to the volume of the product if it is gaseous, provided all the volumes are measured at the same temperature and pressure.* For example, when hydrogen and chlorine combine, 1 dm³ (or litre) of hydrogen will combine exactly with 1 dm³ of chlorine to form 2 dm³ of hydrogen chloride. When nitrogen and hydrogen combine, a certain volume of nitrogen will combine with three times that volume to form twice its volume of ammonia.

This relationship seemed so neat that chemists tried to think of a simple explanation of it. At the same time as Gay Lussac was putting forward his Law in France, the British chemist *John Dalton* was publishing his *Atomic Theory.* One of the things he proposed was that chemical combination takes place between small whole numbers of atoms. For example, one atom of iron combines with one atom of sulphur; two atoms of hydrogen combine with one atom of oxygen. A Swedish chemist called Berzelius tried to explain Gay Lussac's Law in terms of Dalton's Atomic Theory by suggesting that equal volumes of gases contain equal numbers of atoms, but this explanation was not completely satisfactory. The Italian chemist Avogadro gave a better explanation in 1811. His suggestion, known as *Avogadro's Hypothesis,* is that: *Equal volumes of all gases (at the same temperature and pressure) contain the same number of molecules.* This makes it easy to understand Gay Lussac's Law because if

1 volume of hydrogen + 1 volume of chlorine
$\rightarrow$ 2 volumes of hydrogen chloride

then

n molecules of hydrogen + n molecules of chlorine
$\rightarrow$ $2n$ molecules of hydrogen chloride

1 molecule of hydrogen + 1 molecule of chlorine
$\rightarrow$ 2 molecules of hydrogen chloride

Since it must be possible to make 1 molecule of hydrogen chloride, each molecule of hydrogen must contain 2 (or a multiple of 2) atoms, and each molecule of chlorine must contain 2 (or a multiple of 2) atoms, and the above equation becomes:

$$H_2(g) + Cl_2(g) \rightarrow 2HCl(g)$$

It follows from Avogadro's Hypothesis that, whenever we see an equation representing a reaction between gases, we can substitute volumes of gases in the same ratio as numbers of molecules. Thus,

$$N_2(g) + 3H_2(g) \rightarrow 2NH_3(g)$$

means that since

1 molecule of nitrogen + 3 molecules of hydrogen form 2 molecules of ammonia

then 1 volume of nitrogen + 3 volumes of hydrogen form 2 volumes of ammonia.

For example, 1 dm³ of nitrogen + 3 dm³ of hydrogen form 2 dm³ of ammonia.

Since equal volumes of gases (at the same temperature and pressure) contain the same number of molecules, if you consider the Avogadro constant, N molecules of carbon dioxide, N molecules of hydrogen, N molecules of oxygen, and so on, then all these gases will occupy the same volume. The volume occupied by N molecules of gas, which is a mole of each gas, is called the molar volume of a gas. The molar volume of a gas is 22.4 dm³ at 0 °C and 1 atmosphere pressure. These conditions are referred to as standard temperature and pressure (s.t.p.) or as normal temperature and pressure (n.t.p.). Sometimes the gas molar volume is quoted as 24.0 dm³ at room temperature (20 °C) and 1 atmosphere pressure.

The unit cubic decimetres (dm³) is used for gas molar volume here. In the section of questions from GCE papers, you will meet other units. Some Boards use cubic centimetres (cm³), and some use litres (l).

$$1 \text{ dm}^3 = 1\,000 \text{ cm}^3 = 1 \text{ l}$$

Calculations on the reacting volumes of gases are very simple as they depend on the fact that

A mole of gas occupies 22.4 dm³ at standard temperature and pressure
or 24.0 dm³ at room temperature and 1 atmosphere

Example 1 What volume of carbon dioxide (at s.t.p.) is produced by burning 12 g of carbon?

Method: The equation

$$C(s) + O_2(g) \rightarrow CO_2(g)$$

tells us that, 1 atom of carbon forms 1 molecule of carbon dioxide, therefore, 1 mole of carbon forms 1 mole of carbon dioxide

that is,   12 g of carbon form 22.4 dm³ of carbon dioxide at s.t.p.
               (relative atomic mass in g)   (molar volume of a gas)

Answer: 12 g of carbon produce 22.4 dm³ of carbon dioxide.

Example 2 \What volume of hydrogen (at s.t.p.) is evolved when 0.325 g of zinc reacts with dilute hydrochloric acid?

Method: From the equation

$$Zn(s) + 2HCl(aq) \rightarrow H_2(g) + ZnCl_2(aq)$$

we see that 1 mole of zinc gives 1 mole of hydrogen

therefore   65 g of zinc give 22.4 dm³ of hydrogen at s.t.p.

        1 g of zinc gives 22.4/65 dm³ of hydrogen at s.t.p.

        0.325 g of zinc gives

$$\frac{22.4 \times 0.325}{65} \text{ dm}^3 \text{ of hydrogen at s.t.p.}$$

$$= 0.112 \text{ dm}^3 \text{ or } 112 \text{ cm}^3 \text{ of hydrogen}$$

Answer: 112 cm³ of hydrogen.

Example 3 What volume of oxygen is required for the complete combustion of 125 cm³ of butene? What volume of carbon dioxide is formed? (All volumes are quoted at the same temperature and pressure.)

Method: The equation

$$C_4H_8(g) + 6O_2(g) \rightarrow 4CO_2(g) + 4H_2O(g)$$

shows that

1 mole of butene + 6 moles of oxygen form 4 moles of carbon dioxide;
1 volume of butene + 6 volumes of oxygen form 4 volumes of carbon dioxide;
and 125 cm³ of butene + 750 cm³ of oxygen form 500 cm³ of carbon dioxide.

Answer: Volume of oxygen = 750 cm³. Volume of carbon dioxide = 500 cm³.

Using the Reacting Volumes of Gases to work out the Equation for a Reaction

From the volumes of gases taking part in a reaction can be worked out the number of moles of each gas taking part in a reaction. From this the equation follows.

Example 1   100 cm³ of propane burn in 500 cm³ of oxygen to form 300 cm³ of carbon dioxide and 400 cm³ of steam. What is the equation for the reaction?

Method: Since

100 cm³ propane + 500 cm³ oxygen → 300 cm³ CO_2 + 400 cm³ steam,

| 1 molecule of propane | + 5 molecules of oxygen | 3 molecules of CO_2 | + 4 molecules of steam |

and the equation is, therefore,

Answer: $C_3H_8(g) + 5O_2(g) → 3CO_2(g) + 4H_2O(g)$

Example 2   200 cm³ of ammonia burn in 250 cm³ of oxygen to form 200 cm³ of nitrogen monoxide and 300 cm³ of steam. What is the equation for this reaction?

Method: Since

| 200 cm³ ammonia | + 250 cm³ oxygen → 200 cm³ NO | + 300 cm³ steam, |

| 2 volumes of ammonia | + 2.5 volumes of oxygen | → 2 volumes of nitrogen monoxide | + 3 volumes of steam |

| 2 molecules of NH_3 | + 2.5 molecules of O_2 | → 2 molecules of NO | + 3 molecules of H_2O |

To give an equation with integral numbers of molecules, we multiply through by two, and write

| 4 molecules of NH_3 | + 5 molecules of O_2 | → 4 molecules of NO | + 6 molecules of H_2O |

Answer: $4NH_3(g) + 5O_2(g) → 4NO(g) + 6H_2O(g)$

Calculating the Relative Molecular Mass of a Gas

You know that the molar volume of a gas is 22.4 dm³ at s.t.p. If you know the volume occupied by a known mass of gas, you can calculate the mass required to fill 22.4 dm³ at s.t.p. This will be the molar mass of the gas (the relative molecular mass in grams).

Example 1   If 4.5 g of liquid A vaporise to give 1.12 dm³ of vapour at s.t.p., what is the relative molecular mass of A?

Method:

If 1.12 dm³ are occupied by 4.5 g of A,

$$1 \quad \text{dm}^3 \text{ is occupied by } \frac{4.5}{1.12} \text{ g of A}$$

$$22.4 \text{ dm}^3 \text{ are occupied by } \frac{22.4 \times 4.5}{1.12} \text{ g of A}$$

$$= 20 \times 4.5 = 90 \text{ g of A}$$

Answer: The relative molecular mass of A is 90.

Example 2   1.1 g of B evaporate to give 560 cm³ of vapour at s.t.p. What is the relative molecular mass of B?

Method:

If 560 cm³ are occupied by 1.1 g of B,

$$1 \quad \text{dm}^3 \text{ is occupied by } \frac{1.1}{0.56} \text{ g of B}$$

$$22.4 \text{ dm}^3 \text{ are occupied by } \frac{22.4 \times 1.1}{0.56} \text{ g}$$

$$= 40 \times 1.1 = 44 \text{ g of B}$$

Answer: The relative molecular mass of B is 44.

Problems on Reacting Volumes of Gases at Standard Temperature and Pressure

Section 1

1.   The complete combustion of carbon in oxygen yields carbon dioxide. Calculate the volume of carbon dioxide (at s.t.p.) that would be formed by the combustion of 12 g of carbon.

   First, write the equation for the combustion of carbon.
   How many moles of carbon dioxide are formed from 1 mole of carbon?
   What volume of carbon dioxide is formed from 1 mole of carbon?
   How many moles of carbon are there in 12 g of carbon?
   What volume of carbon dioxide will this number of moles of carbon produce?

2.   What volume of hydrogen is formed when 12 g of magnesium react with an excess of acid?

   First, write the equation for the reaction of magnesium with any dilute acid.

How many moles of hydrogen are produced from 1 mole of magnesium?

What is the volume of this number of moles of hydrogen?

How many moles of magnesium are there in 12 g of the metal?

What is the volume of hydrogen produced from this number of moles of magnesium?

3. What volume of hydrogen is produced when 6.5 g of zinc react with an excess of acid?

First, write the equation for the reaction of zinc with a dilute acid.

How many moles of hydrogen are formed from 1 mole of zinc?

What is the volume of this number of moles of hydrogen?

How many moles of zinc are present in 6.5 g of zinc?

What is the volume of hydrogen formed from this number of moles of zinc?

4. What is the volume of carbon dioxide obtained by heating 10 g of calcium carbonate?

First, write the equation for the thermal decomposition of calcium carbonate.

How many moles of carbon dioxide are formed from 1 mole of calcium carbonate?

What is the volume of this number of moles of carbon dioxide?

How many moles of calcium carbonate are there in 10 g of this solid?

What volume of carbon dioxide is formed from this number of moles?

5. Calculate the volume of oxygen (at s.t.p.) needed for the complete combustion of 125 cm^3 of methane. What volume of carbon dioxide is formed?

First, write the equation for the combustion of methane, CH_4.

How many moles of oxygen react with 1 mole of methane?

How many volumes of oxygen react with 1 volume of methane?

What volume of oxygen reacts with 125 cm^3 of methane?

How many moles of carbon dioxide are formed by combustion of 1 mole of methane?

How many volumes of carbon dioxide are formed from 1 volume of methane?

What volume of carbon dioxide is formed from 125 cm^3 of methane?

6. After an electric spark is passed through a mixture of 250 cm^3 of hydrogen and 250 cm^3 of oxygen, steam is formed. The steam is condensed as water. What is the volume of gas remaining in the apparatus?

First, write the equation.

Does the reaction need equal volumes of hydrogen and oxygen?

Which gas will be left over after the reaction?
Calculate what volume of this gas is used during the reaction.
Subtract the volume used from the starting volume to find out what volume is left.

Section 2

1. The complete combustion of carbon in oxygen yields carbon dioxide. Calculate the volume of oxygen at s.t.p. that would react with 10 g of carbon and the volume of carbon dioxide formed.

2. In the reaction between marble and hydrochloric acid, the equation is

$$CaCO_3(s) + 2HCl(aq) \rightarrow CaCl_2(aq) + CO_2(g) + H_2O(l)$$

What mass of marble would be needed to give 10.00 g of carbon dioxide?
What volume would this gas occupy at s.t.p.?

3. Zinc reacts with aqueous hydrochloric acid to give hydrogen.

$$Zn(s) + 2HCl(aq) \rightarrow H_2(g) + ZnCl_2(aq)$$

What mass of zinc would be needed to give 100 g of hydrogen? What volume would this gas occupy (a) at s.t.p. and (b) at room temperature and 1 atmosphere?

4. Lead(II) oxide can be reduced by hydrogen to lead. What volume of hydrogen at s.t.p. would be needed to reduce 4.46 g of lead(II) oxide to lead? What mass of lead would be formed?

$$PbO(s) + H_2(g) \rightarrow Pb(s) + H_2O(g)$$

5. What volume of oxygen (at s.t.p.) would be formed by the complete thermal decomposition of 3.31 g of lead(II) nitrate? The equation for the reaction is

$$2Pb(NO_3)_2(s) \rightarrow 2PbO(s) + 4NO_2(g) + O_2(g)$$

6. What volume of oxygen (at s.t.p.) is formed by the decomposition of a solution containing 1.7 g of hydrogen peroxide?

$$2H_2O_2(aq) \rightarrow 2H_2O(l) + O_2(g)$$

7. What mass of potassium chlorate(V) must be heated to give 112 cm³ of oxygen (at s.t.p.), according to the equation

$$2KClO_3(s) \rightarrow 2KCl(s) + 3O_2(g)$$

8. A mixture of hydrogen and oxygen can be exploded to form water. What volume of oxygen is needed to convert 123 cm³ of hydrogen into water?

9. Propane burns in oxygen to form carbon dioxide and water:

$$C_3H_8(g) + 5O_2(g) \rightarrow 3CO_2(g) + 4H_2O(g)$$

What volume of oxygen (at s.t.p.) is required for the complete combustion of 44 g of propane? What volume of carbon dioxide is formed?

10. Sodium hydrogencarbonate decomposes on heating, with evolution of carbon dioxide:

$$2NaHCO_3(s) \rightarrow Na_2CO_3(s) + CO_2(g) + H_2O(g)$$

What volume of carbon dioxide (at s.t.p.) can be obtained by heating 4.20 g of sodium hydrogencarbonate? If 4.2 g of sodium hydrogen-carbonate react with an excess of dilute hydrochloric acid, what volume of carbon dioxide (at s.t.p.) is evolved?

11. Calculate the volume of hydrogen (at s.t.p.) required to reduce 250 cm³ of propene, C_3H_6, to propane, C_3H_8.

12. Calculate the volume of carbon dioxide (at s.t.p.) formed by the complete combustion of 250 cm³ of butane, C_4H_{10} (measured at s.t.p.).

13. Calculate the relative molecular masses (or relative formula masses) of the following gases from the information given below:

2.2 g of the gas, A, occupy 1.12 dm³ at s.t.p.

8.0 g of the gas, B, occupy 2.8 dm³ at s.t.p.

4.0 g of the gas, C, occupy 560 cm³ at s.t.p.

1.0 g of the gas, D, occupies 1.4 dm³ at s.t.p.

4.0 g of the gas, E, occupy 3.2 dm³ at s.t.p.

1.7 g of the gas, F, occupy 2.24 dm³ at s.t.p.

14. A compound, A, has empirical formula C_2H_6O. 4.6 g of A vaporise to give 2.24 dm³ of vapour (at s.t.p.). What is the molecular formula of A?

15. The empirical formula of liquid B is C_2H_4O. If 4.4 g of B vaporise to give 1.12 dm³ of vapour, what is the molecular formula of B?

Correction of Gas Volumes to Standard Temperature and Pressure

The volume of a gas increases with a rise in temperature, and decreases with an increase in pressure. When we state the volume of a gas, we must therefore state the temperature and pressure at which it was measured. It is usual to give the volume at 0 °C and 1 atmosphere. These conditions are called standard temperature and pressure (s.t.p.) or normal temperature and pressure (n.t.p.).

It is very seldom that gases are collected at s.t.p. The volume of the gas must be measured, the temperature and pressure noted, and the following method used to work out the volume which the gas would occupy at s.t.p. The method is based on research into the behaviour of gases by two scientists called *Boyle* and *Charles*.

Boyle studied the effect of altering the pressure on the volume of a gas. His Law states: *The pressure of a fixed mass of gas at a constant temperature is inversely proportional to its volume.*

Charles studied the effect of temperature on the volume of a gas. His Law states that: *A fixed mass of gas at constant pressure expands by 1/273 of its volume at 0 °C for every °C rise in temperature.*

Boyle's Law and Charles' Law may be combined to form the *General Gas Equation,* which is

$$\frac{P_1 \times V_1}{T_1} = \frac{P_2 \times V_2}{T_2}$$

Where P_1, V_1 and T_1 are the pressure, volume and temperature (Kelvin) of a fixed mass of gas, and V_2 its volume at pressure P_2 and temperature T_2. This is the equation we use to calculate the volume of a gas under different conditions of temperature and pressure.

Temperatures must be in kelvins. To convert temperatures on the Celsius scale to temperatures on the Kelvin scale, you add 273.

$$\text{Temperature (K)} = \text{Temperature (°C)} + 273$$

$$273 \text{ K} = 0°C$$

The pressure may be in atmospheres (atm), millimetres of mercury (mm Hg) or Pascals (Pa), which have the dimensions Newtons per square metre (N m^{-2}).

$$1 \text{ atm} = 760 \text{ mm Hg} = 1.01 \times 10^5 \text{ Pa} = 1.01 \times 10^5 \text{ N m}^{-2}$$

Volumes can be in cm³ or dm³ (1 dm³ = 1000 cm³).

Note that you must keep *the same units for pressure and volume on both sides of the equation.*

Example 1   A gas is collected at 400 °C and 2 atmospheres pressure. The volume is 125 cm³. Correct this volume to s.t.p.

Method: Use the equation

$$\frac{P_1 \times V_1}{T_1} = \frac{P_2 \times V_2}{T_2}$$

Put the measured values of temperature, pressure and volume into the left-hand side and the standard temperature and pressure into the right-hand side.

The measured values are: $P_1 = 2$ atmospheres, $V_1 = 125$ cm^3, $T_1 = 398$ K.

The standard conditions are: $P_2 = 1$ atmosphere, $T_2 = 273$ K, and V_2 is the unknown volume at s.t.p. Then

$$\frac{2 \times 125}{398} = \frac{1 \times V_2}{273}$$

$$V_2 = \frac{2 \times 125 \times 273}{398}$$

$$= 172 \text{ cm}^3$$

Answer: The volume at s.t.p. is 172 cm^3.

Example 2 A volume of 380 cm^3 of gas was collected at 77 °C and 840 mm mercury. What would be the volume of the gas at s.t.p.?

Method: The measured values are:
$V_1 = 380$ cm^3, $T_1 = 273 + 77 = 350$ K, $P_1 = 840$ mm mercury.
The standard conditions are: $T_2 = 273$ K, $P_2 = 760$ mm mercury.
Note that you must use the same units for pressure on both sides of the equation. Since

$$\frac{P_1 \times V_1}{T_1} = \frac{P_2 \times V_2}{T_2}$$

$$\frac{840 \times 380}{350} = \frac{760 \times V_2}{273}$$

$$V_2 = \frac{840 \times 380 \times 273}{350 \times 760}$$

$$= 328 \text{ cm}^3$$

Answer: The volume at s.t.p. is 328 cm^3.

Example 3 What volume of carbon dioxide, measured at 546 °C and 2.02×10^5 N m^{-2} pressure, will be obtained by the complete thermal decomposition of 10 g of calcium carbonate?

Method: First, write the equation:

$$CaCO_3(s) \rightarrow CaO(s) + CO_2(g)$$

This shows that 1 mole of calcium carbonate forms 1 mole of carbon dioxide.

For a solid, 1 mole = Relative Formula Mass in grams

RFM of $CaCO_3$ = $40 + 12 + (3 \times 16) = 100$

For a gas, 1 mole   = 22.4 dm³ at s.t.p.

Therefore, 1 mole of carbon dioxide = 22.4 dm³ at s.t.p.

From the equation,

100 g of calcium carbonate form 22.4 dm³ of carbon dioxide

and 10 g of calcium carbonate form $\dfrac{10}{100} \times 22.4$

$$= 2.24 \text{ dm}^3 \text{ of carbon dioxide.}$$

To correct this volume to 546 °C and 2.02×10^5 N m⁻² use the expression

$$\frac{P_1 \times V_1}{T_1} = \frac{P_2 \times V_2}{T_2}$$

Substitute $V_1 = 2.24$ dm³, $T_1 = 273$ K, $P_1 = 1.01 \times 10^5$ N m⁻²

$$T_2 = 819 \text{ K}, P_2 = 2.02 \times 10^5 \text{ N m}^{-2}$$

Then $\dfrac{1.01 \times 10^5 \times 2.24}{273} = \dfrac{2.02 \times 10^5 \times V_2}{819}$

$$V_2 = \frac{2.24 \times 819 \times 1.01 \times 10^5}{273 \times 2.02 \times 10^5}$$

$$= 3.36 \text{ dm}^3$$

Answer: 10 g of calcium carbonate give 3.36 dm³ of carbon dioxide at 2 atmospheres and 546 °C.

Alternative Method: Some pupils prefer to tackle these problems in a slightly different manner. In Example 2, we corrected a volume of 380 cm³ at 77 °C and 840 mm mercury to s.t.p. Here is the alternative method:

First, consider the effect of the change in temperature.

As the gas cools from 77 °C (350 K) to 0 °C (273 K), the volume will decrease.

The measured volume must be multiplied by a factor less than 1, by $\dfrac{273}{350}$.

The new volume is $380 \times \dfrac{273}{350}$ cm³.

Secondly, consider the effect of changing the pressure.
As the pressure decreases from 840 mm to 760 mm mercury, the gas expands. The measured volume must be multiplied by a factor greater than 1, by $\dfrac{840}{760}$.

The new volume is $380 \times \dfrac{273}{350} \times \dfrac{840}{760} = 328 \text{ cm}^3$.

You will see that this slightly different way of applying the gas laws gives almost the same answer!

Problems on Correction of Gas Volumes to Standard Temperature and Pressure

Section 1

No calculators or logarithms are needed for this section.

1. A gas occupies 1 dm^3 at 27 °C and 1 atmosphere pressure. What will the new volume be at 0 °C and 0.5 atmosphere?

2. A gas occupies 360 cm^3 at 50 °C and 800 mm mercury. At 0 °C and 760 mm mercury, the volume will be

    (a) $\dfrac{360 \times 760 \times 273}{323 \times 800}$

    (b) $\dfrac{323 \times 760 \times 273}{360 \times 800}$

    (c) $\dfrac{360 \times 800 \times 273}{323 \times 760}$

    (d) $\dfrac{360 \times 800 \times 323}{273 \times 760}$

3. A gas collected at 60 °C and 2 atmospheres occupies 150 cm^3. Will the volume at s.t.p. be closest to
    (a) 150 cm^3 (b) 250 cm^3 (c) 350 cm^3 (d) 450 cm^3?

4. A gas occupies 450 cm^3 at 1 atmosphere and 60 °C. What will be its volume at 1.5 atmospheres and 171 °C?

5. A gas occupies 600 cm^3 at 80 °C and 900 mm mercury. Which of the following shows its volume at 0 °C and 760 mm mercury?

    (a) $\dfrac{760 \times 900 \times 273}{600 \times 353}$

    (b) $\dfrac{600 \times 900 \times 273}{353 \times 760}$

    (c) $\dfrac{600 \times 900 \times 353}{273 \times 760}$

    (d) $\dfrac{760 \times 353}{600 \times 900 \times 273}$

6. At 70 °C and 800 mm mercury, a gas measures 750 cm³. Which of the following expressions shows what will be the volume at 90 °C and 880 mm mercury?

(a) $\dfrac{750 \times 880 \times 363}{343 \times 800}$

(b) $\dfrac{750 \times 800 \times 343}{880 \times 363}$

(c) $\dfrac{750 \times 800 \times 363}{343 \times 880}$

(d) $\dfrac{750 \times 800 \times 880}{363 \times 343}$

7. Correct the following volumes of gas to s.t.p. (0 °C and 760 mm mercury):
   (a) 190 cm³ gas at 27 °C and 800 mm mercury
   (b) 95 cm³ gas at 57 °C and 880 mm mercury
   (c) 380 cm³ gas at 273 °C and 840 mm mercury
   (d) 380 cm³ gas at 96 °C and 738 mm mercury
   (e) 190 cm³ gas at 182 °C and 800 mm mercury.

8. A given mass of gas occupies 1 dm³ at 30 °C and 1 atmosphere pressure. What would be the volume of the gas at 60 °C and 0.5 atmosphere pressure?
   (a) 0.5 dm³   (b) 2 dm³   (c) 4 dm³
   (d) between 2 and 3 dm³   (e) 8 dm³

9. A gas collected at 50 °C and 1 atmosphere pressure occupies 500 cm³. What volume would the gas occupy at 20 °C and 2 atmospheres pressure?
   (a) 250 cm³   (b) 200–240 cm³   (c) 500 cm³
   (d) 1 000 cm³   (e) 800–900 cm³

Section 2

Calculators or logarithm tables are needed for this section.

1. The following volumes of gas were measured at s.t.p. Calculate what the volume would be under the stated conditions.
   (a) 150 cm³ of gas at s.t.p. corrected to 37 °C and 800 mm mercury
   (b) 120 cm³ of gas at s.t.p. corrected to 57 °C and 570 mm mercury
   (c) 250 cm³ of gas at s.t.p. corrected to 100 °C and 5 atmospheres
   (d) 600 cm³ of gas at s.t.p. corrected to 200 °C and 1 000 mm mercury.

2. Calculate the volume of a mole of gas at
   (a) 1.12×10^5 N m^{-2}
   (b) 7.97×10^4 N m^{-2}
   (c) 1.51×10^5 N m^{-2}
   (d) 2.98×10^5 N m^{-2}

3. Calculate the volume of 0.2 mole of gas at
   (a) 800 mm mercury and 27 °C
   (b) 400 mm mercury and 409.5 °C
   (c) 2 atmospheres and 87 °C
   (d) 150 mm mercury and 177 °C

4. (a) Calculate the volume of hydrogen (at 20 °C and 800 mm) formed by the reaction of 0.12 g of magnesium with an excess of dilute acid.

   (b) What volume of carbon dioxide (at 500 °C and 1.5 atmospheres) is formed by the action of 20 g of marble with an excess of dilute hydrochloric acid?

   (c) Calculate the volume of sulphur dioxide which can be collected at 50 °C and 850 mm from the reaction of 21 g of sodium sulphite with an excess of dilute hydrochloric acid, on warming.

$$Na_2SO_3(s) + 2HCl(aq) \rightarrow 2NaCl(aq) + SO_2(g) + H_2O(l)$$

   (d) What volume of oxygen, measured at 67 °C and 800 mm, can be obtained from the thermal decomposition of 245 g of potassium chlorate(V)?

$$2KClO_3(s) \rightarrow 2KCl(s) + 3O_2(g)$$

Section 3   Problems on Reacting Volumes of Gases from GCE O-level Papers

1. 20 cm^3 of ethane and 100 cm^3 of oxygen were exploded and the mixture allowed to attain the original room temperature and pressure. What will be the volume of each of the remaining gases?

$$2C_2H_6 + 7O_2 = 4CO_2 + 6H_2O \qquad \text{(SUJB)}$$

2. 7 g of nitrogen and 11 g of a gaseous hydrocarbon X both have the same volume measured at s.t.p.

   (a) How many moles of nitrogen molecules are there in 7 g of nitrogen?

   (b) What is the relative molecular mass of the hydrocarbon X?

   (O & C)

3. Calculate the volume of oxygen at s.t.p. which theoretically could be obtained from 50 cm³ of a solution of hydrogen peroxide which contains 68 g/litre. (O & C)

4. (a) What is the volume (measured at s.t.p.) of 17.0 g ammonia?

   (b) What is the volume (measured at s.t.p.) of 17.0 g of a mixture of nitrogen and hydrogen in the proportions by volume of one of nitrogen to three of hydrogen? (Oxford)

5. Methane will undergo combustion:

$$CH_4 + 2O_2 \rightarrow CO_2 + 2H_2O$$

   (a) How many moles are present in 64 g of methane?

   (b) How many moles of oxygen are needed to burn completely 64 g of methane? (Oxford)

6. 100 cm³ of nitrogen oxide (NO) were mixed with 50 cm³ of oxygen (O_2). This is a ratio of moles of NO : moles of O_2 =

   (a) 1:1   (b) 1:2   (c) 2:1   (d) 100:1   (e) 1:50

   Choose the correct answer.

   If the two gases reacted according to the equation

$$2NO(g) + O_2(g) \rightarrow N_2O_4(g)$$

   what volume of gas would be formed by the reaction between 100 cm³ of NO and 50 cm³ of O_2?

   (a) 50 cm³   (b) 100 cm³   (c) 150 cm³   (d) 200 cm³   (e) 250 cm³
   (London)

7. The equation for the reaction between the gases carbon monoxide and oxygen is

$$2CO(g) + O_2(g) \rightarrow 2CO_2(g)$$

   If 200 cm³ of carbon monoxide were burnt in 200 cm³ of oxygen (all volumes being measured at the same temperature and pressure), the final volume, in cm³, will be

   (a) 400   (b) 300   (c) 250   (d) 200   (e) 150   (London)

8. Warming a gaseous oxide of chlorine, Cl_xO_y, causes it to decompose:

$$2Cl_xO_y \rightarrow xCl_2 + yO_2$$

   In an experiment, 20 cm³ of the oxide gave 20 cm³ of chlorine and 10 cm³ of oxygen (all gas volumes being measured at the same temperature and pressure). Calculate the formula of this oxide, showing your reasoning clearly. (AEB 1977)

9. Calculate the volume of oxygen which would be used to burn completely 250 cm^3 of butane gas. (Assume that all volumes are measured at the same temperature and pressure.) (O & C)

10. Calculate the number of molecules in 560 cm^3 of propane gas at s.t.p. (1 mole of a gas occupies 22400 cm^3 at s.t.p. The Avogadro constant (number) is $6.02 \times 10^{23} \text{ mol}^{-1}$.)

   Calculate the volume of oxygen that would be used to burn completely 500 cm^3 of propane gas. (Assume that all volumes are measured at the same temperature and pressure.) (O & C)

11. Aqueous hydrogen peroxide contains $34 \text{ g } H_2O_2/l$. Calculate the volume of oxygen, measured at s.t.p., which could be obtained from 10 cm^3 of this solution. (O & C)

12. The equation for the complete combustion of butane, C_4H_{10}, is

$$2C_4H_{10} + 13O_2 = 8CO_2 + 10H_2O$$

   (a) 100 cm^3 of butane at room temperature and pressure are burnt in excess oxygen. Calculate the volume of oxygen used, and the volume of carbon dioxide produced, after adjusting to the original temperature and pressure.

   (b) Calculate:

   (i) the number of molecules in 5600 cm^3 of hydrogen at s.t.p.

   (ii) the number of atoms in 20 g of calcium.

   (One mole of gas occupies 22400 cm^3 at s.t.p. The Avogadro constant (number) is 6.02×10^{23}.) (O & C)

13. 20 cm^3 of ethane, C_2H_6, are exploded with 100 cm^3 of oxygen, both gases being at room temperature and at atmospheric pressure. Calculate the volume which each of the residual gases would occupy after adjusting to the original conditions of temperature and pressure. (O & C)

14. If N is the Avogadro constant, calculate in terms of N

   (a) the number of atoms in 10 g of calcium

   (b) the number of H_2O molecules in 72 g of water

   (c) the number of molecules in 11.2 litres of hydrogen at s.t.p. (O & C)

15. One mole of gas occupies 22400 cm^3 at s.t.p. The Avogadro constant (number) is 6.02×10^{23}.

   (a) 15 cm^3 of methane, CH_4, are exploded with 80 cm^3 of oxygen, both gases being at room temperature and at atmospheric pressure.

Calculate the volume which each of the residual gases would occupy after adjusting to the original conditions of temperature and pressure.

(b) Calculate

    (i) the number of molecules in $11\,200\;cm^3$ of oxygen at s.t.p.

    (ii) the number of atoms in 4.8 g of magnesium.   (O & C)

16. At the same temperature and pressure, 5.60 g of nitrogen occupies the same volume as 8.00 g of argon. Use this information to calculate the relative molecular mass of argon and the number of atoms present in 1 molecule of argon. Mention two ways in which argon differs from nitrogen.   (Oxford)

17. (a) when $100\;cm^3$ of gaseous hydrocarbon, P, was heated in the presence of a catalyst, with $100\;cm^3$ of hydrogen, a single new gaseous compound, Q, was formed, there being no P and no hydrogen left over. Explain clearly what this information tells you about the gas P.

    (b) When $100\;cm^3$ of the gas Q was mixed with an excess of oxygen and an electric spark passed, there was an explosion, $350\;cm^3$ of the oxygen had reacted, and $200\;cm^3$ of carbon dioxide had been formed. The steam, which was the only other product of the explosion, had condensed to a drop of water of negligible volume.

    Explain clearly what this information tells you about the gas Q.

    (c) Give the name and formula of P.

    (d) Describe a simple chemical test which will distinguish between P and Q.   (London)

18. (a) What are the volumes measured at s.t.p. of 8.00 g of each of the following gases: (i) methane, (ii) oxygen, (iii) sulphur dioxide?

    (b) Given that one mole of a gas contains N molecules, what number of molecules is present in the masses of the samples (i), (ii) and (iii) given above?   (Oxford)

19. The equation for the combustion of ethane is

$$2C_2H_6 + 7O_2 \rightarrow 4CO_2 + 6H_2O$$

If 60 g of ethane were completely burned in oxygen,

(a) what volume of carbon dioxide would be formed?

(b) how many moles of oxygen would be needed?

(c) what volume of air would be needed to effect complete combustion of the 60 g of ethane (assuming air contains 20% by volume of oxygen)?

(d) what change in pressure would be needed to halve the volume of carbon dioxide (assuming temperature conditions remained the same)?   (Oxford)

20. 10 cm³ of hydrogen sulphide were burnt with 50 cm³ of oxygen, both gases being at laboratory temperature and pressure. Calculate the volume of each residual gas after it had been adjusted to the original temperature and pressure.   (O & C)

21. The Avogadro constant (number) is 6.02×10^{23}. Calculate the number of molecules in 5 600 cm³ of a gas at s.t.p. What weight of chlorine at s.t.p. would contain this number of molecules?   (O & C)

22. 10 cm³ of carbon monoxide are exploded with 50 cm³ of oxygen. Calculate the volume which each of the remaining gases would occupy when they have been adjusted to the original temperature and pressure.   (O & C)

23. Calculate the maximum volume of oxygen, measured at s.t.p., which could be given off from an aqueous solution of hydrogen peroxide containing 0.34 g H_2O_2.   (O & C)

24. 20 cm³ of carbon monoxide are burnt with 30 cm³ of oxygen. Calculate the volume of each of the residual gases after adjustment to the original temperature and pressure.   (O & C)

25. The equation for the combustion of butane is

$$2C_4H_{10} + 13O_2 + 8CO_2 + 10H_2O$$

What volume of oxygen would be required to burn completely 100 cm³ of butane? Calculate the number of molecules in 5 600 cm³ of butane gas at s.t.p.   (O & C)

26. If 20 cm³ of ethane, C_2H_6, are burned in an excess of oxygen, state the minimum volume of oxygen required for complete combustion and the volume of carbon dioxide formed. Assume that all volumes are measured at the same temperature and pressure.   (JMB)

27. One mole of gas occupies 24 dm³ at room temperature and pressure. The volume of hydrogen produced at room temperature and pressure when 0.2 mole of sodium is treated with an excess of water is

(a) 1.2 dm³ (b) 2.4 dm³ (c) 4.8 dm³ (d) 9.6 dm³ (e) 12 dm³
(Cambridge)

28. 1.4 g of nitrogen has a volume of 1.12 dm³ at s.t.p. Assuming that the relative atomic mass of nitrogen is 14.0 and the molar volume of a gas is 22.4 dm³ at s.t.p., calculate the number of atoms in one molecule of nitrogen.   (Cambridge)

29. In an experiment, 1 cm³ of a hydrocarbon X requires 4 cm³ of oxygen for complete combustion to give 3 cm³ of carbon dioxide, all gas volumes being measured at s.t.p. Which of the following formulae represents X?

    (a) CH_4 (b) C_2H_3 (c) C_2H_4 (d) C_3H_4 (e) C_4H_8

    (Cambridge)

30. 60 cm³ of oxygen can be converted into 40 cm³ of ozone (both volumes being measured under the same conditions of temperature and pressure). State Avogadro's Law, and use this law to calculate the formula of ozone.   (Cambridge)

31. On decomposition under suitable conditions, 10 cm³ of an oxide of nitrogen produced 10 cm³ of nitrogen and 5 cm³ of oxygen. The mass of 1 litre of the gaseous oxide under room conditions was $1^5/_6$ g. Deduce the formula of this oxide of nitrogen.

    $(A_r(N) = 14; A_r(O) = 16;$ molar volume of a gas at room temperature and pressure is 24 litres (dm³).)   (Welsh JEC)

32. An organic liquid has the following percentage composition by mass: carbon 37.5%, hydrogen 12.5%, oxygen 50.0%.

    1.60 g of the liquid on vaporisation produced 1.12 dm³ (l) of the vapour, measured at standard temperature and pressure (s.t.p.).

    Calculate

    (a) the empirical formula

    (b) the relative molecular mass

    (c) the molecular formula of the liquid.   (AEB 1978)

33. On complete combustion, 7 g of a hydrocarbon X, produced 22 g carbon dioxide and 9 g water. 21 g of X had a volume of 11.2 dm³ corrected to standard temperature and pressure.

    Find

    (a) How many grams of hydrogen and carbon are separately present in 7 g of X

    (b) the empirical formula of X

    (c) the relative molecular mass of X

    (d) the molecular formula of X.   (AEB 1979)

Section 4   Problems from GCE O-level Papers involving both Masses of Solids and Volumes of Gases

1. Calcium nitrate decomposes on heating according to the equation

    $$2Ca(NO_3)_2 = 2CaO + 4NO_2 + O_2$$

    The relative molecular mass of calcium nitrate is 164.

    What volume at s.t.p. of (a) nitrogen(IV) oxide (nitrogen dioxide), (b) oxygen is evolved when 16.4 g of calcium nitrate is heated to constant weight? (SUJB)

2. Calculate the volume of ammonia, measured at room temperature and pressure, obtained when 0.1 mol of ammonium sulphate reacts with calcium hydroxide according to the equation

    $$(NH_4)_2SO_4 + Ca(OH)_2 \rightarrow CaSO_4 + 2NH_3 + 2H_2O$$
    (AEB 1977)

3. Calculate (a) the mass, (b) the volume measured at s.t.p. of carbon dioxide released when 4.20 g sodium hydrogencarbonate is decomposed according to the equation

    $$2NaHCO_3 = Na_2CO_3 + H_2O + CO_2 \qquad \text{(Oxford)}$$

4. When solid silver carbonate is gently heated, the following change takes place:

    $$2Ag_2CO_3 \rightarrow 4Ag + 2CO_2 + O_2$$

    0.01 mol of silver carbonate is decomposed in this reaction. Calculate:

    (a) the volume of oxygen obtained (measured at room temperature and pressure)

    (b) the mass of silver that remains.

    ($Ag = 108$; one mole of gas occupies 24 dm³ at room temperature and pressure.) (AEB 1977)

5. (a) If 2 g of hydrogen contain N molecules, write down in terms of N the number of

    (i)   atoms in 2 g of argon

    (ii) molecules in 2 g of bromine

    (iii) molecules in 2 g of nitrogen oxide, NO

    (iv) ions in 2 g of carbonate ions, CO_3^{2-}.

    (b) The Avogadro constant (number) has the value 6×10^{23}. A crystal of sucrose ($C_{12}H_{22}O_{11}$) weighs 0.0342 g. How many molecules does it contain? (London)

6. The decomposition of potassium nitrate by heat is represented by the following equation:

$$2KNO_3 = 2KNO_2 + O_2$$

Calculate the volume of oxygen at s.t.p. evolved when 0.1 mole of nitrate is decomposed. (O & C)

7. What volume of hydrogen is obtained, under room conditions, when 0.13 g of zinc reacts with an excess of dilute sulphuric acid? (The molar volume of a gas is 24 dm^3 under room conditions.)

(Cambridge)

8. On complete combustion, 0.100 mole of one of the compounds (a) to (e) gave 13.2 g of carbon dioxide. Identify the compound, explaining your reasoning.

(a) CH_3OH (b) CH_3COOH (c) $CH_3CH = CH_2$
(d) CH_3CH_2OH (e) $CH_3CH_2CH_2CH_3$

(Cambridge)

6. Calculations on Reacting Volumes of Solutions

Concentration

We measure the concentration of a solution by saying how many moles of solute are dissolved in one cubic decimetre (or litre) of solution. A solution containing 1 mole of solute in 1 dm^3 of solution is often referred to as a *molar* solution, or an M solution.

For example, the relative formula mass of sodium carbonate is 106 ($Na_2CO_3 : (2 \times 23) + 12 + (3 \times 16) = 106$). A solution made by dissolving 106 g of anhydrous sodium carbonate, putting the solution into a 1 dm^3 graduated flask and adding water until the solution just reaches the graduation mark contains 1 mol dm^{-3}, and can be called a molar solution. If the same mass of sodium carbonate were dissolved and made up to 2 dm^3 of solution, the concentration of the solution would be 0.5 mol dm^{-3} (and we could call it a half-molar, or 0.5 M, solution). Dissolved in 500 cm^3 of solution, this mass of sodium carbonate would give a solution with concentration 2 mol dm^{-3} (which we could call a two-molar, or 2 M, solution). If two moles of sodium carbonate were dissolved and made up to 1 dm^3, the solution would have a concentration 2 mol dm^{-3} (and could be referred to as a two-molar solution).

$$\text{Concentration of solution (in mol dm}^{-3}) = \frac{\text{Number of moles of solute}}{\text{Volume of solution in dm}^3}$$

(often referred to as Molarity)

or

$$\text{Number of moles of solute} = \text{Concentration (mol dm}^{-3}) \times \text{Volume (dm}^3)$$

(Molarity of solution)

Figure 6.1 illustrates this.

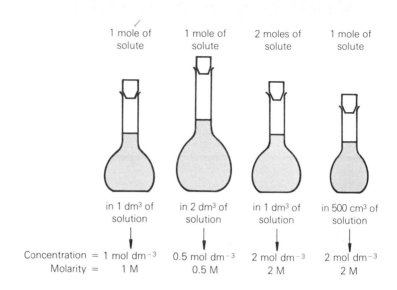

Figure 6.1   How to calculate concentration

Example 1   Calculate the concentration in mol dm⁻³ (molarity) of a solution containing 36.5 g of hydrogen chloride in 4 dm³ of solution.

Method:

Relative formula mass of hydrogen chloride

$$= (35.5 + 1.0) = 36.5$$

Number of moles present in 36.5 g = 1 mole

Volume   = 4 dm³

Concentration (mol dm⁻³) of solution = $\dfrac{\text{Number of moles of solute}}{\text{Volume of solution in dm}^3}$

(or Molarity)

$$= 1/4 = 0.25.$$

Answer: The concentration is 0.25 mol dm⁻³; the solution is 0.25 M.

Example 2   Calculate the number of moles of solute in 250 cm³ of a solution of sodium hydroxide having a concentration 2 mol dm⁻³ (a 2 M solution).

Method:

Concentration (or Molarity) of solution = 2 mol dm^{-3}

Volume of solution   = 250 cm^3 = 0.25 dm^3

Number of moles of solute = Concentration × Volume
                            (mol dm^{-3})   (dm^3)
                            (or Molarity)

                         = 2 × 0.25 = 0.5

Answer: The solution contains 0.50 moles of solute.

Problems on Concentration (or Molarity)

1. Calculate the concentration in mol dm^{-3} (molarity) of
   (a) 3.65 g of hydrogen chloride in 2 dm^3 of solution
   (b) 73.0 g of hydrogen chloride in 2 dm^3 of solution
   (c) 49.0 g of sulphuric acid in 2 dm^3 of solution
   (d) 49.0 g of sulphuric acid in 250 cm^3 of solution
   (e) 2.80 g of potassium hydroxide in 500 cm^3 of solution
   (f) 28.0 g of potassium hydroxide in 4 dm^3 of solution
   (g) 5.30 g of anhydrous sodium carbonate in 200 cm^3 of solution
   (h) 53.0 g of anhydrous sodium carbonate in 2.5 dm^3 of solution

2. Calculate the number of moles of solute in
   (a) 250 cm^3 of sodium hydroxide solution containing 1 mol dm^{-3}
   (b) 500 cm^3 of sodium hydroxide solution containing 0.25 mol dm^{-3}
   (c) 250 cm^3 of 0.02 M calcium hydroxide solution (0.02 mol dm^{-3})
   (d) 2 dm^3 of 1.25 M sulphuric acid (1.25 mol dm^{-3})
   (e) 125 cm^3 of aqueous nitric acid, having a concentration of 0.40 mol dm^{-3}
   (f) 200 cm^3 of ammonia solution, having a concentration of 0.125 mol dm^{-3}
   (g) 123 cm^3 of aqueous hydrochloric acid of concentration 3 mol dm^{-3}
   (h) 1500 cm^3 of potassium hydroxide solution of concentration 0.75 mol dm^{-3}.

Neutralisation

When hydrochloric acid is neutralised by sodium hydroxide, the equation is

$$HCl(aq) + NaOH(aq) \rightarrow NaCl(aq) + H_2O(l)$$

1 mole of hydrochloric acid needs 1 mole of sodium hydroxide.
1 mole of HCl is present in 1 dm^3 of solution of concentration 1 mol dm^{-3}.
1 mole of NaOH is present in 1 dm^3 of solution of concentration 1 mol dm^{-3}.
1 dm^3 of hydrochloric acid of concentration 1 mol dm^{-3} therefore neutralises
1 dm^3 of sodium hydroxide of concentration 1 mol dm^{-3}.

Does 1 dm^3 of acid of concentration 1 mol dm^{-3} always neutralise 1 dm^3 of alkali of concentration 1 mol dm^{-3}? If you consider the reactions

$$HNO_3(aq) + KOH(aq) \rightarrow KNO_3(aq) + H_2O(l)$$

and

$$H_2SO_4(aq) + Na_2CO_3(aq) \rightarrow Na_2SO_4(aq) + CO_2(g) + H_2O(l)$$

you will find that 1 dm^3 of acid of concentration 1 mol dm^{-3} neutralises 1 dm^3 of alkali of concentration 1 mol dm^{-3}.

However, if you consider the neutralisation of sulphuric acid by sodium hydroxide solution

$$H_2SO_4(aq) + 2NaOH(aq) \rightarrow Na_2SO_4(aq) + 2H_2O(l)$$

1 mole of sulphuric acid neutralises 2 moles of sodium hydroxide, and, therefore, 1 dm^3 of sulphuric acid of concentration 1 mol dm^{-3} neutralises 2 dm^{-3} of sodium hydroxide solution of concentration 1 mol dm^{-3}.

In the reaction between hydrochloric acid and sodium carbonate solution

$$2HCl(aq) + Na_2CO_3(aq) \rightarrow 2NaCl(aq) + CO_2(g) + H_2O(l)$$

2 moles of hydrochloric acid are needed by 1 mole of sodium carbonate, and 2 dm^3 of hydrochloric acid of concentration 1 mol dm^{-3} neutralise 1 dm^3 of sodium carbonate solution of concentration 1 mol dm^{-3}.

When you are doing calculations on reacting volumes of solutions, the equation for the reaction is important. Look at the equation to find out how many moles of acid react with one mole of alkali. The problems in Section 1 on page 71 will give you practice at this.

Titration

It is possible to find out the concentration (or molarity) of a solution by finding out what volume of it will react with a known volume of a solution of known concentration. For example, you can find out the concentration of an alkaline solution by finding out what volume of it will neutralise, say, 25 cm³ of a solution of an acid of known concentration. The procedure of adding one liquid to another in a measured way is called *titration*. A solution of known concentration (or molarity) is called a *standard solution*.

Example 1   In a titration of a solution of unknown concentration sodium hydroxide against standard hydrochloric acid, 25.0 cm³ of sodium hydroxide solution neutralise 22.5 cm³ of aqueous hydrochloric acid of concentration 0.9 mol dm⁻³. What is the concentration in mol dm⁻³ of the sodium hydroxide solution?

Method: In tackling this calculation,

(a) Find out the number of moles of acid needed to neutralise one mole of alkali.

(b) Use the expression

Number of moles of solute = Concentration × Volume
(mol dm⁻³)   (dm³)

Number of moles of hydrochloric = Concentration × Volume
acid   (mol dm⁻³)   (dm³)

$$= 0.9 \times 22.5 \times 10^{-3}$$

Since

$$NaOH(aq) + HCl(aq) \rightarrow NaCl(aq) + H_2O(l)$$

Number of moles of hydrochloric = Number of moles of sodium
acid   hydroxide

Therefore, number of moles of $= 0.9 \times 22.5 \times 10^{-3}$
sodium hydroxide

But, number of moles of sodium = Volume × Concentration
hydroxide   (dm³)   (mol dm⁻³)

$$= 25.0 \times 10^{-3} \times M$$

where M = concentration in mol dm⁻³

$$25.0 \times 10^{-3} \times M = 0.9 \times 22.5 \times 10^{-3}$$

$$M = \frac{0.9 \times 22.5 \times 10^{-3}}{25.0 \times 10^{-3}}$$

$$= 0.81$$

Answer: The sodium hydroxide solution contains 0.81 mol dm^{-3}. (It is 0.81 M.)

Example 2 Titration shows that 25.0 cm^3 of sodium hydroxide solution neutralise 22.5 cm^3 of sulphuric acid of concentration 0.9 mol dm^{-3} (0.9 M). What is the concentration of the sodium hydroxide solution?

Method: As in Example 1:
(a) Find the number of moles of acid needed to neutralise 1 mole of alkali.

(b) Number of moles of solute = Concentration of solution × Volume (dm^3)

Number of moles of sulphuric acid	= Volume × Concentration (dm^3) (mol dm^{-3})

$$= 22.5 \times 10^{-3} \times 0.9$$

Since

$$2NaOH(aq) + H_2SO_4(aq) \rightarrow Na_2SO_4(aq) + 2H_2O(l)$$

Number of moles of sodium hydroxide = 2 × number of moles of sulphuric acid

$$= 2 \times 22.5 \times 10^{-3} \times 0.9$$

But, number of moles of sodium hydroxide = Volume × Concentration

$$= 25.0 \times 10^{-3} \times M$$

where M = concentration in mol dm^{-3}

Therefore, $25.0 \times 10^{-3} \times M = 2 \times 22.5 \times 10^{-3} \times 0.9$

$$M = \frac{2 \times 22.5 \times 10^{-3} \times 0.9}{25.0 \times 10^{-3}}$$

$$= 1.62$$

Answer: The concentration of the sodium hydroxide solution is 1.62 mol dm^{-3}.

Note that, although the titration figures in this example are the same as those in Example 1, the concentration of sodium hydroxide is twice what it was calculated to be in Example 1 because, in Example 2, 1 mole of acid neutralises *2* moles of alkali.

Example 3   If 25.0 cm³ of calcium hydroxide solution neutralise 35.0 cm³ of aqueous hydrochloric acid of concentration 0.08 mol dm⁻³ (0.08 M), what is the concentration of the calcium hydroxide solution?

Method:

Number of moles of hydrochloric = Volume × Concentration
acid

$$= 35.0 \times 10^{-3} \times 0.08$$

Since

$$Ca(OH)_2(aq) + 2HCl(aq) \rightarrow CaCl_2(aq) + 2H_2O(l)$$

Number of moles of calcium   = ½ × number of moles of
hydroxide   hydrochloric acid

$$= 0.5 \times 35.0 \times 10^{-3} \times 0.08$$

But, number of moles of calcium = Volume × Concentration
hydroxide

Let M = concentration;

then moles $Ca(OH)_2 = 25.0 \times 10^{-3} \times M$

$$25.0 \times 10^{-3} \times M = 0.5 \times 35.0 \times 10^{-3} \times 0.08$$

$$M = \frac{0.5 \times 35.0 \times 10^{-3} \times 0.08}{25.0 \times 10^{-3}}$$

$$= 0.056$$

Answer: The solution has a concentration of 0.056 mol dm⁻³. (It is 0.056 M.)

Example 4   What volume of hydrochloric acid of concentration 0.25 mol dm⁻³ (0.25 M) is needed to neutralise 5.3 g of anhydrous sodium carbonate?

Method: You tackle this example, as before, by finding the number of moles of each substance. For the solid, use the expression:

Number of moles = Mass/Relative formula mass.

For the solution, use the expression:

Number of moles of solute $\quad$ = Volume $\times$ Concentration
$\qquad\qquad\qquad\qquad\qquad\quad$ (dm³) $\qquad$ (mol dm⁻³)

Relative formula mass of $\qquad$ = $(2 \times 23) + 12 + (3 \times 16) = 106$
$\quad$ Na₂CO₃

Number of moles of Na₂CO₃ $\quad$ = Mass/RFM = 5.3/106

$\qquad\qquad\qquad\qquad\qquad\qquad\quad$ = 1/20 = 0.05 mol

Since

$$Na_2CO_3(s) + 2HCl(aq) \rightarrow 2NaCl(aq) + CO_2(g) + H_2O(l)$$

Number of moles of HCl $\qquad$ = 2 $\times$ number of moles of Na₂CO₃

$\qquad\qquad\qquad\qquad\qquad\qquad\quad$ = 2×0.05 = 0.1 mol

But number of moles of HCl in $\quad$ = Volume $\times$ Concentration
$\quad$ solution $\qquad\qquad\qquad\qquad\qquad$ (dm³) $\qquad$ (mol dm⁻³)

Therefore $\qquad\qquad\qquad\qquad\quad$ 0.1 = Volume $\times$ 0.25

$\qquad\qquad\qquad\qquad$ Volume = 0.1/0.25 = 0.4 dm³

Answer: The volume needed is 0.4 dm³ or 400 cm³.

Example 5 What volume of hydrochloric acid of concentration 2.00 mol dm⁻³ is required to react with 2.50 g of calcium carbonate?

Method: This problem follows the same pattern as Example 4.

Number of moles of calcium $\quad$ = Mass/Relative formula mass
$\quad$ carbonate

Relative formula mass of CaCO₃ = $40 + 12 + (3 \times 16)$ = 100

Therefore, number of moles of $\quad$ = 2.50/100 = 0.025
$\quad$ CaCO₃

Since

$$CaCO_3(s) + 2HCl(aq) \rightarrow CaCl_2(aq) + CO_2(g) + H_2O(l)$$

1 mole of calcium carbonate reacts with 2 moles of hydrochloric acid, and number of moles of
$\quad$ hydrochloric acid $\qquad\qquad\qquad$ = 2×0.025 = 0.050.

But number of moles of $\qquad\qquad$ = Volume $\times$ Concentration
$\quad$ hydrochloric acid $\qquad\qquad\qquad$ (dm³) $\qquad$ (mol dm⁻³)

Therefore $$0.050 = \text{Volume} \times 2.00$$
$$\text{Volume} = 0.050/2.00$$
$$= 0.025 \text{ dm}^3$$

Answer: The volume of hydrochloric acid is 0.025 dm³ or 25 cm³.

Example 6 A sample of ammonium chloride is warmed with 250 cm³ of sodium hydroxide of concentration 1 mol dm⁻³ until the evolution of ammonia ceases. The excess of sodium hydroxide is neutralised by 130 cm³ of sulphuric acid of concentration 0.50 mol dm⁻³. What mass of ammonium chloride was present?

Method: This example looks more difficult as there are two steps in it:

(1) the reaction between ammonium chloride and alkali:

$$NH_4Cl(s) + NaOH(aq) \rightarrow NH_3(g) + NaCl(aq) + H_2O(l)$$

(2) the neutralisation of excess alkali by sulphuric acid:

$$2NaOH(aq) + H_2SO_4(aq) \rightarrow Na_2SO_4(aq) + 2H_2O(l)$$

You tackle it in the same way as the other examples: calculate the number of moles of the substances for which you have the necessary information. You can find the number of moles of sulphuric acid used as you know the volume and the concentration. The number you obtain will tell you the number of moles of sodium hydroxide left over after reaction (1).

You can find the number of moles of sodium hydroxide added in stage (1) as you are told the volume and concentration of sodium hydroxide solution used. The difference between the number of moles added and the number of moles left over will give the number of moles of sodium hydroxide used in reaction (1). This will tell you the number of moles of ammonium chloride with which it reacted.

Number of moles of sulphuric acid	= Volume × Concentration (dm³) (mol dm⁻³)
	= 130 × 10⁻³ × 0.5 = 0.065
Number of moles of sodium hydroxide left over	= 2 × 0.065 = 0.13
Number of moles of sodium hydroxide added	= Volume (dm³) × Concentration
	= 250 × 10⁻³ × 1 = 0.25

Number of moles of sodium
hydroxide used in reaction (1) $= 0.25 - 0.13 = 0.12$

Therefore, number of moles of
ammonium chloride $= 0.12$

Relative formula mass of $NH_4Cl = 14 + (4 \times 1) + 35.5 = 53.5$

Mass of ammonium chloride $=$ Number of moles $\times$ Relative
formula mass

$= 0.12 \times 53.5 = 6.42$ g

Answer: The sample contained 6.42 g of ammonium chloride.

Problems on Reacting Volumes of Solutions

Section 1

Calculators and logarithms are not needed for these problems.

The following are problems on neutralisation. Show, giving your
working, whether each of these statements is true or false.

1. 1 mole of HNO_3 will neutralise
    (a) 5 dm^3 of KOH(aq) of concentration 0.2 mol dm^{-3}: True or False?
    (b) 2 dm^3 of NaOH(aq) of concentration 0.2 mol dm^{-3}
    (c) 2 dm^3 of KOH(aq) of concentration 0.5 mol dm^{-3}
    (d) 0.5 dm^3 of NaOH(aq) of concentration 1 mol dm^{-3}
    (e) 250 cm^3 of Na_2CO_3(aq) of concentration 2 mol dm^{-3}
    (f) 200 cm^3 of Na_2CO_3(aq) of concentration 4 mol dm^{-3}

2. 1 mole of H_2SO_4 will neutralise
    (a) 500 cm^3 of NaOH(aq) of concentration 4 mol dm^{-3}: True or False?
    (b) 1 dm^3 of KOH(aq) of concentration 1 mol dm^{-3}
    (c) 400 cm^3 of NaOH(aq) of concentration 5 mol dm^{-3}
    (d) 500 cm^3 of Na_2CO_3(aq) of concentration 1 mol dm^{-3}
    (e) 2 dm^3 of Na_2CO_3(aq) of concentration 0.5 mol dm^{-3}
    (f) 4 dm^3 of KOH(aq) of concentration 0.25 mol dm^{-3}

3. 4 moles of HCl will neutralise
    (a) 4 dm^3 of NaOH(aq) of concentration 0.5 mol dm^{-3}: True or
    False?
    (b) 4 dm^3 of Na_2CO_3(aq) of concentration 0.5 mol dm^{-3}

(c) 2 dm³ of Na_2CO_3(aq) of concentration 1 mol dm⁻³

(d) 500 cm³ of KOH(aq) of concentration 5 mol dm⁻³

(e) 200 cm³ of NaOH(aq) of concentration 10 mol dm⁻³

(f) 2.5 dm³ of KOH(aq) of concentration 1.6 mol dm⁻³

4. 2 moles of KOH will neutralise
   (a) 1 dm³ of HCl(aq) of concentration 2 mol dm⁻³: True or False?
   (b) 250 cm³ of H_2SO_4(aq) of concentration 4 mol dm⁻³
   (c) 200 cm³ of HNO_3(aq) of concentration 5 mol dm⁻³
   (d) 250 cm³ of H_2SO_4(aq) of concentration 2 mol dm⁻³
   (e) 500 cm³ of HCl(aq) of concentration 4 mol dm⁻³
   (f) 250 cm³ of HNO_3(aq) of concentration 5 mol dm⁻³

5. 5 moles of NaOH will neutralise
   (a) 2 dm³ of HCl(aq) of concentration 2 mol dm⁻³: True or False?
   (b) 250 cm³ of HCl(aq) of concentration 10 mol dm⁻³
   (c) 250 cm³ of H_2SO_4(aq) of concentration 10 mol dm⁻³
   (d) 500 cm³ of H_2SO_4(aq) of concentration 5 mol dm⁻³
   (e) 2 500 cm³ of HNO_3(aq) of concentration 2 mol dm⁻³
   (f) 2 dm³ of HNO_3(aq) of concentration 2 mol dm⁻³

6. 0.5 mole of Na_2CO_3 will neutralise
   (a) 1 dm³ of HCl(aq) of concentration 1 mol dm⁻³: True or False?
   (b) 1 dm³ of H_2SO_4(aq) of concentration 1 mol dm⁻³
   (c) 500 cm³ of HCl(aq) of concentration 1 mol dm⁻³
   (d) 250 cm³ of HNO_3(aq) of concentration 2 mol dm⁻³
   (e) 200 cm³ of H_2SO_4(aq) of concentration 2.5 mol dm⁻³
   (f) 500 cm³ of HNO_3(aq) of concentration 2 mol dm⁻³

Section 2

These problems can be solved without the use of calculators.

1. 25.0 cm³ of a solution of potassium hydroxide are neutralised by 35.0 cm³ of aqueous hydrochloric acid of concentration 0.75 mol dm⁻³. Calculate the concentration in mol dm⁻³ (molarity) of the potassium hydroxide solution.

2. If 25.0 cm³ of a solution of sodium hydroxide are neutralised by 27.5 cm³ of aqueous sulphuric acid of concentration 0.25 mol dm⁻³, what is the concentration (mol dm⁻³) of the sodium hydroxide solution?

3. A mixture of sodium chloride and sodium carbonate weighing 10.0 g was dissolved in water and titrated against hydrochloric acid. For neutralisation, 40 cm³ of aqueous hydrochloric acid of concentration 1 mol dm⁻³ (i.e. molar) were needed. Calculate the masses of sodium chloride and sodium carbonate in the mixture.

4. A solution of sodium hydroxide contains 8.0 g dm⁻³. 25.0 cm³ of this solution neutralise 35.0 cm³ of a solution of nitric acid.
   (a) Calculate the concentration of the sodium hydroxide solution in mol dm⁻³.
   (b) Calculate the concentration of the nitric acid in mol dm⁻³.
   (c) Calculate the concentration of nitric acid in g dm⁻³.

5. What is the volume of sulphuric acid containing 1.5 mol dm⁻³ which would be needed to react with 7.95 g of copper(II) oxide?

6. Sodium reacts with water according to the equation

$$2Na(s) + 2H_2O(l) \rightarrow 2NaOH(aq) + H_2(g)$$

   Find the volume of aqueous hydrochloric acid, of concentration 0.25 mol dm⁻³, needed to neutralise the sodium hydroxide formed by the reaction of 0.23 g of sodium.

7. A solution of hydrochloric acid was titrated against a solution of sodium carbonate of concentration 0.25 mol dm⁻³ (i.e. 0.25 M). 25.0 cm³ of the acid solution neutralised 20.0 cm³ of the alkali. Calculate
   (a) the number of moles of alkali used in the titration
   (b) the number of moles of acid used in the titration
   (c) the concentration of the acid in mol dm⁻³.

8. 2.1 g of sodium hydrogencarbonate react with dilute hydrochloric acid with the evolution of carbon dioxide. What volume of a solution of hydrochloric acid containing 0.50 mol dm⁻³ is needed to liberate the maximum possible volume of carbon dioxide? What volume of carbon dioxide (at s.t.p.) is obtained?

9. A solution of sodium carbonate contains 53.0 g in 250 cm³ of solution. Calculate
   (a) the concentration (mol dm⁻³) of the sodium carbonate solution
   (b) the volume of aqueous hydrochloric acid of concentration 0.25 mol dm⁻³ (i.e. 0.25 M) needed to neutralise 25.0 cm³ of the sodium carbonate solution.

10. Hydrogen is formed by the reaction of a solution of sulphuric acid with an excess of magnesium. What volume of aqueous sulphuric acid, of concentration 1.5 mol dm^{-3}, must be used to give 1 120 cm^3 of hydrogen (at s.t.p.)?

Section 3   Questions from GCE O-level Papers

You will notice that the abbreviation M for molar, meaning 'having a concentration 1 mol dm^{-3}, appears in a number of questions.

1. Hydrogen peroxide solution can be prepared in the laboratory by the reaction of barium peroxide with sulphuric acid:

$$BaO_2(s) + H_2SO_4(aq) \rightarrow BaSO_4(s) + H_2O_2(aq)$$

Calculate the mass of barium peroxide needed to react with 100 cm^3 of aqueous sulphuric acid of concentration 0.5 mol dm^{-3}.

(AEB 1978)

2. Calculate the volume of hydrochloric acid of concentration 1 mol dm^{-3} required to neutralise 0.1 mol of sodium carbonate by the reaction

$$Na_2CO_3 + 2HCl \rightarrow 2NaCl + CO_2 + H_2O$$

(AEB 1977)

3. The formula of crystals of sodium sulphate(VI) is $Na_2SO_4 \cdot 10H_2O$. Sodium sulphate can be made by the reaction

$$2NaOH + H_2SO_4 = Na_2SO_4 + 2H_2O$$

What is (a) the relative molecular mass of the crystals

(b) the percentage of water of crystallization in the crystals

(c) the volume of M sulphuric acid required to give 16.1 g of crystals? (SUJB)

4. What volume of carbon dioxide, measured at s.t.p., is evolved when 200 g of calcium carbonate is heated to constant mass, and what minimum volume of 2 M aqueous sodium hydroxide is required to absorb the carbon dioxide evolved?

$$NaOH + CO_2 = NaHCO_3$$ (SUJB)

5. What minimum volume of 2 M sulphuric acid would be required to dissolve 4.05 g of zinc oxide? (SUJB)

6. Scandium (symbol Sc) reacts with hydrochloric acid releasing hydrogen:

$$2Sc + 6HCl \rightarrow 2ScCl_3 + 3H_2$$

It was found that 3 g of scandium reacted exactly with 20 cm^3 of aqueous hydrochloric acid of concentration 10 mol dm^{-3}. Calculate the mass of scandium which reacts with 6 mol of hydrochloric acid. Then use the equation to deduce the relative atomic mass (atomic) weight) of scandium. (AEB 1977)

7. (a) Describe the preparation of reasonably dry crystals of magnesium sulphate ($MgSO_4 \cdot 7H_2O$), starting from magnesium carbonate. What mass in grams of the crystals could you expect from 0.1 mole of magnesium carbonate?

(b) 25.0 cm^3 of a solution of sodium carbonate is neutralised by 20 cm^3 of a solution of sulphuric acid containing 0.1 mole/litre. What is the concentration of the sodium carbonate in mole/litre and in grams of Na_2CO_3/litre? Describe how this experiment could be carried out. (O & C)

8. (a) Write balanced equations for the reaction of potassium hydroxide solution with dilute sulphuric acid to form (i) the acid salt, (ii) the normal salt.

(b) What volume of potassium hydroxide solution of concentration 0.1 mol/l, will need to be added to 50 cm^3 of dilute sulphuric acid of concentration 0.05 mol/l to form (i) the acid salt, (ii) the normal salt? (Welsh JEC)

9. In an experiment to determine the percentage of calcium carbonate in a sample of limestone, it was found that 1 g of the limestone neutralised 38 cm^3 of hydrochloric acid of concentration 0.5 mol/l.

(a) Write an equation for the reaction between calcium carbonate and hydrochloric acid.

(b) What fraction of a mole of hydrochloric acid was used?

(c) What fraction of a mole of calcium carbonate is required to neutralise the hydrochloric acid?

(d) What mass of calcium carbonate does this represent?

(e) Calculate the percentage of calcium carbonate in the limestone. (Welsh JEC)

10. Excess barium chloride solution is added to 50 cm^3 of sulphuric acid which contains 0.1 mol/l. Calculate the mass of dry barium sulphate which could be obtained. (O & C)

11. (a) When a solution of hydrochloric acid was titrated against 0.90 M sodium hydroxide solution, 23.0 cm^3 of the acid was needed to neutralise 25.0 cm^3 of the alkali. Calculate
(i) the number of moles of alkali used in each titration

    (ii) the number of moles of acid used in the titration

    (iii) the number of moles of acid in 1.00 dm³ of solution

    (iv) the concentration of the acid in g/dm³.

  (b) If 0.90 M sodium carbonate solution had been used in (a) instead of 0.90 M sodium hydroxide solution, what volume of the same acid would have been needed to neutralise completely 25.0 cm³ of the sodium carbonate solution?   (Oxford)

12. The reaction between sodium and water is represented by the equation

$$2Na + 2H_2O \rightarrow 2NaOH + H_2$$

4.60 g of sodium are completely dissolved in water. Calculate

(a) the mass of hydrogen produced

(b) the number of moles of hydrogen produced

(c) the volume of hydrogen at s.t.p. produced

(d) the volume of 1.00 M hydrochloric acid needed to neutralise the sodium hydroxide solution produced.   (Oxford)

13. 2.50 g of pure calcium carbonate were completely dissolved in 100 cm³ of 1.00 M nitric acid.

(a) Write an equation for the reaction.

(b) Calculate the number of moles of nitric acid which have reacted.

(c) How many moles of nitric acid remain?

(d) 25.0 cm³ of a solution of potassium hydroxide were needed to neutralise the unused nitric acid. What was the concentration of the potassium hydroxide solution in moles per dm³?

(e) If 0.5 M sodium carbonate solution had been used instead of potassium hydroxide, what volume of solution would have been needed for complete neutralisation?   (Oxford)

14. 50.0 cm³ of a solution containing 9.70 g of the acid H_3NSO_3 per dm³ were exactly neutralised by 62.5 cm³ of a solution of sodium hydroxide containing 0.08 moles NaOH per dm³. Calculate

(a) the volume of alkali neutralising 1 000 cm³ of acid

(b) the volume of alkali neutralising 1 mole of the acid

(c) the number of moles of NaOH neutralising 1 mole of acid

(d) Write the equation for the reaction.   (Oxford)

15. What do you understand by the term mole as it applies to

(a) magnesium, Mg

(b) the volume of a gas at s.t.p.

(c) sulphuric acid, H_2SO_4?

1.20 g magnesium were added to 75.0 cm^3 of a solution of sulphuric acid whose concentration was 1.00 mol dm^{-3}. When reaction ceased, the excess acid was exactly neutralised by adding 25.0 cm^3 of 2.00 M sodium hydroxide solution. From these results calculate

(d) the number of moles of H_2SO_4 reacting with 1 mole of magnesium

(e) the volume of hydrogen at s.t.p. liberated in the course of the experiment. (Oxford)

16. 1 dm^3 of a solution M is made containing 24.0 g of a mixture of sodium hydroxide and sodium chloride. When 25.0 cm^3 of solution M are titrated against hydrochloric acid, 20.8 cm^3 of the acid are required for exact neutralisation; the acid contains 21.9 g HCl per dm^3. Calculate

(a) the concentration of sodium hydroxide in moles of NaOH per dm^3 of solution M

(b) the percentage by mass of sodium chloride present in the original mixture. (Oxford)

17. 0.1 M hydrochloric acid was run from a burette into a flask containing 25 cm^3 of aqueous sodium hydroxide until neutralisation took place. 20 cm^3 of acid were required.

(a) Write down an equation for the reaction that takes place in the flask.

(b) What volume of the acid contains 1 mole of HCl?

(c) How many moles of HCl were neutralised in the experiment?

(d) How many moles of NaOH were present in 25 cm^3 of the aqueous sodium hydroxide?

(e) What was the molarity of the aqueous sodium hydroxide?

(London)

18. (a) Solution X contains 20 g of sodium hydroxide, NaOH, per 250 cm^3 of solution. Calculate the molarity of the solution.

(b) Solution Y contains 180 g of a solid acid, which may be represented by the formula H_nA, per 1 000 cm^3 of solution. The formula weight of the acid is 90. Calculate the molarity of the solution.

(c) 40 cm^3 of solution X are found to react completely with 20 cm^3 of solution Y.

    (i) Calculate the number of moles of NaOH in 40 cm^3 of solution X.

    (ii) Calculate the number of moles of H_nA in 20 cm^3 of solution Y.

(iii) How many moles of NaOH react with one mole of H_nA?

(iv) What is the value of n in H_nA?

(v) Write an equation for the reaction which has taken place.

(London)

19. 25 cm³ of a solution containing 6.0 g of sodium hydroxide per litre exactly neutralise 30 cm³ of a solution of nitric acid.

   (a) Calculate the molarity of the sodium hydroxide solution.

   (b) Using the above results, calculate the molarity of the nitric acid and its concentration in grams per litre.   (JMB)

20. Describe carefully how you would determine the molarity of dilute aqueous sodium hydroxide, using hydrochloric acid of known molarity.

   If the alkali contained 1.50 g of sodium hydroxide in 250 cm³ of solution, and the acid was 0.1 molar, what volume of acid would be needed to react with 20 cm³ of alkali?   (London)

21. (a) Calculate the mass of residue left after heating (to constant mass) 9.33 g of a sample of sodium hydrogencarbonate containing 10% by mass of sodium chloride.

   (b) Calculate the volume, in cm³, of 1.00 mol/dm³ sulphuric acid solution that would just neutralise the residue in (a).

   (The thermal decomposition of sodium hydrogencarbonate may be represented by the equation

   $$2NaHCO_3 \rightarrow Na_2CO_3 + H_2O + CO_2)$$

   (Cambridge)

22. You are required to make 100 cm³ of a solution 0.1 mol/dm³ with respect to Cu^{2+} ions, using copper(II) sulphate pentahydrate $CuSO_4 \cdot 5H_2O$.

   How much of the salt should you weigh out?

   (a) 0.64 g   (b) 1.60 g   (c) 1.78 g   (d) 2.50 g   (e) 6.40 g

   (Cambridge)

7. Electrolysis

An electrolyte is a compound which, molten or in solution, will conduct electricity. The passage of electricity splits the compound up into simpler substances, a process called electrolysis. The vessel in which electrolysis is carried out is called a cell, and the conductors by which the current enters and leaves the cell are called electrodes. Electrolytes are ionic compounds, and, at the electrodes, ions are discharged and new substances are formed.

Deposition of Metals

When a current passes through a solution of a salt of a metal low in the electrochemical series, the metal ions are discharged, and metal atoms are deposited on the cathode (the negative electrode). The electrolysis of a solution of a silver salt, to form a layer of silver on the cathode, is carried out as shown in Figure 7.1. The cathode process is

$$Ag^+(aq) + e^- \rightarrow Ag(s)$$

This equation tells us that 1 silver atom accepts 1 electron to form 1 atom of silver, and therefore 1 mole of silver ions accept 1 mole of electrons to form 1 mole of silver atoms. We can tell when 1 mole of silver atoms (6×10^{23} atoms) have been deposited on the cathode because we know that 1 mole of silver atoms weigh 108 g (the

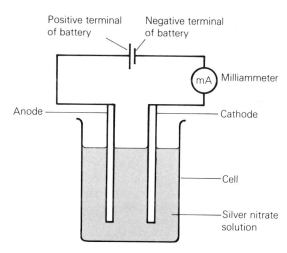

Figure 7.1 Electrolysis cell and circuit

relative atomic mass of silver in grams). As 1 mole of electrons must have passed through the cell to deposit 1 mole of silver atoms, if we measure the quantity of electrical charge which passes through the cell when 108 g of silver are deposited, we shall know that this quantity of electrical charge is equal to 1 mole of electrons. Electrical charge is measured in coulombs.

One coulomb of charge = 1 ampere of current flowing for 1 second
Number of coulombs = Number of amperes × Time in seconds:
$$C = A \times s$$

Experiments can be done in which a known current, measured by means of a milliammeter in the circuit, is passed for a known time through a solution of a silver salt. By weighing the cathode before and after the passsage of the current, the mass of silver deposited can be found. Such experiments show that 96 500 coulombs are required to deposit 108 g of silver. This quantity of electricity must therefore be the charge on a mole of electrons (6×10^{23} electrons). The ratio 96 500 coulombs per mole ($C\ mol^{-1}$) is called the *Faraday constant,* after the famous electrochemist, Michael Faraday.

(Until recently, it was the custom to refer to 96 500 coulombs or a mole of electrons as a *Faraday* of charge. You will see that some questions from past examination papers use this unit of charge.)

Consider the deposition of copper during the electrolysis of a solution of a copper salt:
$$Cu^{2+}(aq) + 2e^- \rightarrow Cu(s)$$
1 copper ion needs 2 electrons to become 1 atom of copper, and
1 mole of copper ions need 2 moles of electrons to become 1 mole of copper atoms.
1 mole of copper ions need $2 \times 96\ 500$ coulombs to become 1 mole of copper atoms.

When gold is deposited during the electrolysis of gold(III) salts, the electrode process is
$$Au^{3+}(aq) + 3e^- \rightarrow Au(s)$$
1 gold ion accepts 3 electrons to form 1 atom of gold.
1 mole of gold ions accept 3 moles of electrons to form 1 mole of gold atoms.
$3 \times 96\ 500$ coulombs are needed to deposit 1 mole of gold.

Calculations in electrochemistry are as easy as one, two, three, once you have worked out whether:

1 mole of electrons discharge 1 mole of the element, e.g. silver, or
2 moles of electrons discharge 1 mole of the element, e.g. copper, or
3 moles of electrons discharge 1 mole of the element, e.g. gold.

No. of moles of element discharged $=\dfrac{\text{No. of moles of electrons}}{\text{No. of charges on one ion of the element}}$

$=\dfrac{\text{No. of coulombs}/96\,500}{\text{No. of charges on one ion of the element}}$

Calculation of the mass of an element deposited by electrolysis

Example 1 What masses of the following elements are deposited by the passage of one mole of electrons through solutions of their salts:

(a) silver, (b) copper, (c) gold?

Method: As already reasoned, since

$$Ag^+(aq) + e^- \rightarrow Ag(s)$$

1 mole of electrons deposits 1 mole of silver = 108 g silver. Since

$$Cu^{2+}(aq) + 2e^- \rightarrow Cu(s)$$

1 mole of electrons deposits ½ a mole of copper = ½ × 63.5 = 31.75 g copper. Since

$$Au^{3+}(aq) + 3e^- \rightarrow Au(s)$$

1 mole of electrons deposits ⅓ mole of gold = ⅓ × 197 = 65.7 g gold.

Answer: (a) 108 g silver, (b) 31.75 g copper, (c) 65.7 g gold

Example 2

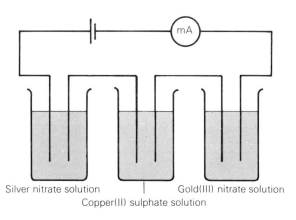

Silver nitrate solution   Gold(III) nitrate solution
Copper(II) sulphate solution

In the apparatus illustrated, a milliammeter registers 10 milliamps for 4 hours. What mass of metal is deposited at each cathode?

Method:

Since number of coulombs = amperes × seconds,

Number of coulombs = $0.01 \times 4 \times 60 \times 60 = 144$ coulombs

Since 1 mole of electrons = 96 500 coulombs,

Electrical charge passed = 144 / 96 500 moles of electrons

Since

$$Ag^+(aq) + e^- \rightarrow Ag(s)$$

1 mole of electrons will deposit 1 mole of silver;

therefore 144 / 96 500 moles of electrons will deposit 144 / 96 500 moles of silver

$$= \frac{144 \times 108}{96\,500} \text{ g silver}$$

$$= 0.161 \text{ g silver}$$

Since

$$Cu^{2+}(aq) + 2e^- \rightarrow Cu(s)$$

1 mole of electrons will deposit ½ mole of copper;

therefore 144 / 96 500 moles of electrons will deposit ½ × 144 / 96 500 moles of copper

$$= \frac{1 \times 144 \times 63.5}{2 \times 96\,500} \text{ g}$$

$$= 0.0474 \text{ g copper}$$

Since

$$Au^{3+}(aq) + 3e^- \rightarrow Au(s)$$

1 mole of electrons will deposit ⅓ mole of gold;

therefore 144 / 96 500 moles of electrons will deposit ⅓ × 144 / 96 500 moles of gold

$$= \frac{1 \times 144 \times 197}{3 \times 96\,500} \text{ g}$$

$$= 0.0980 \text{ g gold.}$$

Answer: 0.161 g silver; 0.0474 g copper; 0.0980 g gold.

Example 3 Aluminium is extracted from its ore by the electrolysis of molten aluminium oxide. How many coulombs of electricity are needed to produce 1 kg of aluminium?

Method: Since

$$Al^{3+}(l) + 3e^- \rightarrow Al(l)$$

3 moles of electrons are needed to give 1 mole of aluminium

$$= 27 \text{ g aluminium}$$

therefore

3/27 moles of electrons are needed to give 1 g aluminium

$1\,000 \times 3/27$ moles of electrons are needed to give 1 kg aluminium

$1\,000 \times 3/27$ moles of electrons $= 1000 \times 3/27 \times 96\,500$ coulombs

$$= 1.07 \times 10^7 \text{ coulombs.}$$

Answer: 1.07×10^7 coulombs will deposit 1 kg aluminium.

Example 4 When a current of 0.156 amperes is passed through a solution of lead(II) nitrate, lead is deposited on the cathode. How long will it take to deposit 0.10 g of lead?

Method: Since

$$Pb^2(aq) + 2e^- \rightarrow Pb(s)$$

2 moles of electrons are needed to deposit 1 mole of lead
$$= 207 \text{ g lead}$$
therefore

2/207 moles of electrons deposit 1 g lead

and

$2/207 \times 0.10$ moles of electrons deposit 0.10 g lead,

$2/207 \times 0.10 \times 96\,500$ coulombs are needed to deposit 0.10 g lead.

Since number of coulombs = amperes × seconds,

$$= 0.0156 \times \text{time in seconds}$$

$$0.0156 \times \text{time} = 2/207 \times 0.10 \times 96\,500$$

$$\text{time} = \frac{2 \times 0.10 \times 96\,500}{207 \times 0.0156}$$

$$= 6000 \text{ seconds}$$

$$= 100 \text{ minutes}$$

$$= 1 \text{ hour } 40 \text{ minutes.}$$

Answer: The current must pass for 1 hour 40 minutes to deposit 0.10 g of lead.

Example 5 A metal of relative atomic mass 27 is deposited by electrolysis. If 0.201 g of the metal is deposited on the cathode when 0.2 amperes flow for three hours, what is the charge on the cations of this metal?

Method: Number of coulombs $= 0.2 \times 3 \times 60 \times 60 = 2160$

If 2160 coulombs deposit 0.201 g of metal,

then 96 500 coulombs deposit $\dfrac{96\,500 \times 0.201}{2160} = 8.98$ g

Since 8.98 g of metal are discharged by 1 mole of electrons (96 500 coulombs),

27 g of metal are discharged by $\dfrac{27}{8.98} \times 1$ mole of electrons

$$= 3 \text{ moles of electrons.}$$

The charge on the metal ions is $+3$.

Answer: The charge on the metal ions is $+3$.

Evolution of Gases

Calculation of the volume of gas evolved during electrolysis

Hydrogen When a current is passed through a solution of a salt of a metal which is high in the electrochemical series, hydrogen ions are discharged at the cathode instead of metal ions. The cathode process is

$$H^+(aq) + e^- \rightarrow H(g)$$

followed by

$$2H(g) \rightarrow H_2(g)$$

Since each hydrogen ion needs 1 electron for its discharge, each hydrogen molecule needs 2 electrons for its formation, and 1 mole of hydrogen needs 2 moles of electrons for its formation.

Thus, 2 moles of electrons will result in the evolution of 1 mole of hydrogen, which has mass 2 g (the relative molecular mass in grams) and volume 22.4 dm^3 (or litres) at s.t.p. (the molar volume of a gas).

Chlorine   When chlorine is evolved at the anode, the anode process is

$$Cl^-(aq) \rightarrow Cl(g) + e^-$$

followed by

$$2Cl(g) \rightarrow Cl_2(g)$$

One chlorine molecule needs 2 electrons for its formation, and 1 mole of chlorine needs 2 moles of electrons for its formation.

Thus, 2 moles of electrons passed through a solution of a chloride will result in the evolution of 71 g of chlorine (the relative molecular mass in grams), which will occupy 22.4 dm^3 at s.t.p. (the molar volume of a gas).

Oxygen   When oxygen is evolved during the electrolysis of a solution, the anode process is

$$OH^-(aq) \rightarrow OH(aq) + e^-$$

followed by

$$4OH(aq) \rightarrow O_2(g) + 2H_2O(l)$$

Thus, 4 hydroxide ions must be discharged to make 1 molecule of oxygen, and 4 moles of electrons must pass to give 1 mole of oxygen (mass 32 g, volume 22.4 dm^3 at s.t.p.).

When oxygen is evolved during the electrolysis of a molten oxide, for example during the extraction of aluminium by the electrolysis of alumina, the anode process is

$$O^{2-}(l) \rightarrow O(g) + 2e^-$$

followed by

$$2O(g) \rightarrow O_2(g)$$

4 electrons are released for the formation of 1 molecule of oxygen, and 4 moles of electrons are required for the formation of 1 mole of oxygen gas.

Example 1 State the names and calculate the volumes of gases formed at the cathode and anode when 15 milliamperes of current are passed for 6 hours through a solution of sulphuric acid.

Method: At the cathode, hydrogen is evolved:

$$H^+(aq) + e^- \rightarrow H(g)$$

$$2H(g) \rightarrow H_2(g)$$

Thus 2 moles of electrons discharge 1 mole of hydrogen molecules. At the anode, oxygen is evolved:

$$OH^-(aq) \rightarrow OH(aq) + e^-$$

$$2OH(aq) \rightarrow H_2O_2(aq)$$

$$2H_2O_2(aq) \rightarrow 2H_2O(l) + O_2(g)$$

showing that 4 moles of electrons are needed to discharge 1 mole of oxygen gas.

Number of coulombs	$= $ amperes $\times$ time in seconds
	$= 15 \times 10^{-3} \times 6 \times 60 \times 60$
	$= 324$ coulombs
Number of moles of electrons	$= 324 / 96\,500$

Since 2 moles of electrons discharge 1 mole of hydrogen molecules,

$$324/96\,500 \text{ moles of electrons discharge} \quad \frac{324}{2 \times 96\,500} \text{ mole of hydrogen}$$

$$= \frac{324 \times 22.4}{2 \times 96\,500} \text{ dm}^3 \text{ of hydrogen}$$

$$= 0.0376 \text{ dm}^3 \text{ or } 37.6 \text{ cm}^3 \text{ hydrogen}$$

Since 4 moles of electrons discharge 1 mole of oxygen molecules, the volume of oxygen evolved will be half that of hydrogen, that is 18.8 cm³.

Answer: At the cathode, 37.6 cm³ of hydrogen are evolved; at the anode, 18.8 cm³ of oxygen are formed.

Problems on Electrolysis

Section 1 No logarithm tables or calculators are needed for this section.

1. Calculate the mass of each of the following elements discharged when 0.25 mole of electrons passes through the solution mentioned:

   (a) copper from copper(II) sulphate solution

   (b) lead from lead(II) nitrate solution

   (c) tin from tin(II) nitrate solution

   (d) bromine from potassium bromide solution

   (e) hydrogen from dilute sulphuric acid

   (f) oxygen from dilute sulphuric acid

2. The relative atomic masses and the symbols of the ions are given for several elements. Calculate the mass of each element deposited when 96 500 coulombs of electricity pass through a molten salt containing the ions.

   (a) rubidium, Rb^+ (RAM = 85.5)

   (b) beryllium, Be^{2+} (RAM = 9)

   (c) strontium, Sr^{2+} (RAM = 87.5)

   (d) lithium, Li (RAM = 7)

   (e) gallium, Ga^{3+} (RAM = 69.5)

   (f) nitrogen, N^{3-} (RAM = 14)

   (g) arsenic, As^{3+} (RAM = 75.0)

   (h) oxygen, O^{2-} (RAM = 16)

3. If 0.50 mole of electrons deposits 29.75 g of tin in electrolysis, what mass of tin will be deposited by 2 moles of electrons?

4. A current of electricity passes through two cells in series. One contains silver nitrate solution, and the other contains lead nitrate solution. In the first, 0.54 g of silver are deposited on the cathode. What mass of lead is deposited in the second cell?

5. A current passes through a solution of copper(II) sulphate, and 0.635 g of copper is deposited on the cathode. What volume of oxygen is evolved at the anode?

Section 2 No calculations or logarithm tables are needed for this section.

1. A current of 1 ampere is passed for 16 minutes, and 0.32 g of copper is deposited on the cathode. The Faraday constant is 96 500

coulombs mole^{-1}. How many coulombs passed through the cell? How many moles of copper were deposited? Calculate the number of moles of electrons needed to deposit one mole of copper atoms from aqueous solution.

2. Which of these expressions shows the mass of aluminium liberated by a current of 0.1 ampere flowing for 6 minutes through molten alumina?

(a) $\dfrac{27 \times 6 \times 60 \times 0.1}{3 \times 96\,500}$

(b) $\dfrac{3 \times 6 \times 60 \times 0.1}{27 \times 96\,500}$

(c) $\dfrac{27 \times 6 \times 60 \times 0.1}{96\,500}$

(d) $\dfrac{27 \times 6 \times 60}{3 \times 96\,500 \times 0.1}$

3. A current of 0.15 ampere flowing for 4 hours deposits 0.71 g copper on a cathode. Which expression gives the charge on a copper ion?

(a) $\dfrac{0.15 \times 4 \times 60 \times 60 \times 63.5}{96\,500 \times 0.71}$

(b) $\dfrac{0.15 \times 4 \times 60 \times 60 \times 0.71}{96\,500 \times 63.5}$

(c) $\dfrac{63.5 \times 0.15 \times 4 \times 60 \times 60 \times 0.71}{96\,500}$

(d) $\dfrac{96\,500 \times 0.71}{0.15 \times 4 \times 60 \times 60 \times 63.5}$

4. A current of 0.1 ampere passes for 3 hours through a solution of dilute sulphuric acid. Which of the following represents the volume of oxygen, in dm^3 at s.t.p., evolved at the anode?

(a) $\dfrac{0.1 \times 3 \times 60 \times 60 \times 4}{96\,500 \times 22.4}$

(b) $\dfrac{0.1 \times 3 \times 60 \times 60 \times 22.4}{96\,500 \times 4}$

(c) $\dfrac{0.1 \times 3 \times 60 \times 60 \times 22.4}{96\,500}$

(d) $\dfrac{0.1 \times 3 \times 60 \times 60}{22.4 \times 96\,500 \times 4}$

5. The mass of lead deposited at the cathode by a current of 0.15 ampere flowing for three hours through a solution of lead(II) nitrate is given by one of the following expressions. Which is correct?

(a) $\dfrac{207 \times 0.15 \times 3 \times 60 \times 60}{96\,500}$

(b) $\dfrac{96\,500 \times 207 \times 2}{0.15 \times 3 \times 60 \times 60}$

(c) $\dfrac{207 \times 0.15 \times 3 \times 60 \times 60}{96\,500}$

(d) $\dfrac{207 \times 0.15 \times 3 \times 60 \times 60}{2 \times 96\,500}$

6. A current of 0.2 ampere passing for 5 hours through a solution of gold ions deposits a mass of 2.45 g of gold on the cathode. Which of these expressions gives the charge on a gold ion?

(a) $\dfrac{2.45 \times 0.2 \times 5 \times 60 \times 60}{197 \times 96\ 500}$

(b) $\dfrac{0.2 \times 5 \times 60 \times 60 \times 197}{96\ 500 \times 2.45}$

(c) $\dfrac{2.45 \times 96\ 500}{197 \times 0.2 \times 5 \times 60 \times 60}$

(d) $\dfrac{197 \times 0.2 \times 5 \times 60 \times 60 \times 96\ 500}{2.45}$

7. A current passes through a solution of sodium chloride, and the hydrogen and chlorine evolved are collected. If 24 000 coulombs of charge flow, what is the mass of hydrogen evolved and its volume? What is the mass of chlorine evolved and its volume? (Faraday constant = 96 000 C mol^{-1})

8. A current passes through two cells in series. One contains silver nitrate solution, and in it 0.216 g of silver is deposited. The other contains cadmium nitrate solution, and in it 0.112 5 g of cadmium is deposited. The relative atomic masses are Ag = 108, and Cd = 112.5. Use this information to calculate the charge on a cadmium ion.

9. How many moles of a trivalent metal are deposited on the cathode during an electrolysis when 2 amperes pass for 96½ hours? If the relative atomic mass of the metal is 27, what is the mass of metal deposited?

10. The Faraday constant is 96 500 coulombs (mole of electrons)$^{-1}$. The relative atomic mass of copper is 64. Which of the following shows the mass of copper liberated at the cathode by a current of 0.1 ampere flowing for 500 seconds through a solution containing copper(II) ions?

(a) $\dfrac{64 \times 500 \times 0.1}{2 \times 96\ 500}$

(b) $\dfrac{64 \times 96\ 500}{2 \times 500 \times 0.1}$

(c) $64 \times 500 \times 0.1 \times 96\ 500$

(d) $\dfrac{64 \times 96\ 500}{500 \times 0.1}$

(e) $\dfrac{64 \times 500 \times 0.1}{96\ 500 \times 60}$

11. Which of the following gives the number of moles of sodium atoms, zinc atoms and aluminium atoms which can be formed from their ions, by passing 96 500 coulombs of electricity?

(a) 1 mol Na, 2 mol Zn, 3 mol Al

(b) 96 500 mol Na, $\dfrac{96\ 500}{2}$ mol Zn, $\dfrac{96\ 500}{3} \times$ mol Al

(c) $\dfrac{1}{96\ 500}$ mol Na, $\dfrac{2}{96\ 500}$ mol Zn, $\dfrac{3}{96\ 500}$ mol Al

(d) 3 mol Na, 2 mol Zn, 1 mol Al

(e) 1 mol Na, ½ mol Zn, ⅓ mol Al

Section 3   Problems from GCE O-level Papers

1. One Faraday of electricity is passed through acidified water. What volume of gas is evolved at each named electrode? (All measurements are at s.t.p.)   (SUJB)

2. A crystalline compound, containing two elements only, has a high melting point. It conducts electricity only if molten, or in aqueous solution.

    (a) Name one such compound, and give its formula.

    (b) Give one reason why the melting point is high.

    (c) At which electrode is each element liberated during the electrolysis of the molten compound?

    (d) How many Faradays are required to liberate 1 mole of atoms of each element? Write the equation for the reaction at each electrode. (F = 96 500 mol⁻¹)   (SUJB)

3. Pure copper can be deposited on an electrode of a copper voltameter. Describe, with the aid of a labelled diagram, how you would do this. Write the equations for the reactions at the electrodes.

    (a) How many Faradays and

    (b) what mass of zinc is required to deposit all the copper from 1 dm³ of 0.5 M aqueous copper(II) ions?

    (c) What is the mass of copper deposited?

    (d) Give an appropriate ionic equation.   (SUJB)

4. When a current of 1 A was passed through aqueous copper sulphate for 50 minutes, 1 g of copper was deposited.

    (a) Calculate the mass of copper which would be deposited by 96 000 C of charge.

    (b) How many coulombs of charge are required to deposit 1 mole of copper atoms?

    (c) Using your answer in (a), deduce the charge on a copper ion.   (AEB 1978)

5. Electric charge equivalent to 1 mole of electrons (1 Faraday) is passed through fused sodium chloride, using carbon electrodes.

   (a) Explain what happens at the anode and state the mass (in grams) of the product.

   (b) Explain what happens at the cathode and state the mass (in grams) of the product. (O & C)

6. 0.01 Faraday of electricity is passed through two cells in series. One cell contains aqueous copper(II) sulphate and the other aqueous silver nitrate. Calculate the weight of the metal which is deposited on the cathode of each cell. (O & C)

7. Two cells A and B are connected in series to a supply of electricity and 0.02 Faradays of electricity are passed through each cell (see the figure below). (N.B. One Faraday = one mole of electrons.) Each cell contains an aqueous solution of copper(II) sulphate of the same concentration, but cell A has copper electrodes and cell B has platinum electrodes.

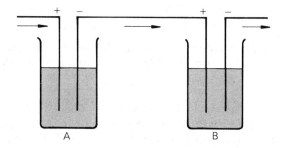

   (a) What happens at (i) the anode in A, (ii) the cathode in B?

   (b) If the cathode in A has increased in mass by 0.63 g, calculate the mass which would have been deposited by one Faraday of electricity at A.

   (c) If the relative atomic mass of the element deposited at the cathode in A is 63.5, calculate the valency of the element deposited.

   (d) (i)   How many moles of oxygen are liberated at the anode in B?
       (ii) What is the mass of this oxygen?
       (iii) What is the volume at s.t.p. of this mass of gas?
                                                              (Oxford)

8. Give equations for the electrode reactions which occur when dilute sulphuric acid is electrolysed using electrodes of platinum. Calculate the volumes, measured at s.t.p., of gases liberated when 4.00 Faradays of charge (4.00 moles of electrons) is passed. (Oxford)

9. If one Faraday of charge (one mole of electrons) will liberate 1.00 g of hydrogen, what charge is required to deposit 15.875 g of copper during electrolysis? (Oxford)

10. Aqueous copper(II) sulphate was electrolysed with copper electrodes. A current of 2 amperes was passed for 8 minutes. During this time the mass of the copper cathode (negative electrode) increased by 0.32 g.
    (a) How many coulombs were passed during the experiment?
    (b) Calculate the number of moles of copper atoms deposited during the experiment.
    (c) Calculate the number of Faradays (moles of electrons) needed to deposit one mole of copper atoms from aqueous solution.
    (d) Write an equation for the reaction at the cathode during electrolysis. (London)

11. A pupil electrolysed a solution of copper(II) sulphate in series with dilute sulphuric acid. A constant current was passed for 30 minutes, and 24 cm³ of hydrogen were collected at the cathode of the dilute sulphuric acid cell.
    (a) The maximum number of cm³ of oxygen evolved at the anode in the dilute sulphuric acid electrolysis would have been
    (i) 6   (ii) 12   (iii) 24   (iv) 32   (v) 48
    (b) If the pupil had doubled the original steady current and allowed it to pass for 15 minutes rather than 30 minutes, then the number of cm³ of hydrogen liberated at the cathode during the electrolysis would have been
    (i) 6   (ii) 12   (iii) 24   (iv) 48   (v) 96
    Choose the correct answer. (London)

12. A metal has only one valency and its relative atomic mass is 27. The formula for its sulphate is $M_2(SO_4)_3$. The number of moles of electrons required to deposit 5.4 g of the metal from a solution of one of its salts is
    (a) 0.2   (b) 0.4   (c) 0.6   (d) 2.7   (e) 5.4
    (Cambridge)

13. Which one of the following would require the largest quantity of electricity for discharge at an electrode?
    (a) 1 mole of Zn^{2+} ions   (b) 2 moles of Fe^{3+} ions
    (c) 3 moles of OH^- ions   (d) 4 moles of Cl^- ions
    (e) 5 moles of Ag^+ ions
    (Cambridge)

14. The number of atoms in one mole (i.e. 12 g) of carbon is 6×10^{23}. Write down

    (a) the number of molecules in 9 g of water

    (b) the number of electrons required to deposit 414 g of lead from fused lead(II) bromide. (Cambridge)

15. How many moles of electrons are required to liberate, by electrolysis, 22.4 dm^3 of oxygen at s.t.p.?

    (a) 1   (b) 2   (c) 4   (d) 8   (e) 16   (Cambridge)

16. Calculate the number of coulombs required to liberate

    (a) 54 g of aluminium   (b) 54 g of silver

    The Faraday constant = 96 500 coulombs/mole.   (Cambridge)

17. Given that 1 mole of a substance contains 6×10^{23} particles, write down:

    (a) the number of molecules in 3.4 g of ammonia

    (b) the mass of copper that can be deposited from copper(II) sulphate solution by 2×10^{23} electrons. (Cambridge)

8. Heat Changes in Chemical Reactions

Heat Changes

There are two types of chemical reactions. In one, the products contain less energy than the reactants; in the other, the products contain more energy than the reactants. The two types of reactions can be represented by the energy diagrams shown in Figure 8.1. The difference between the energy of the reactants and the energy of the products is called the heat of reaction and is represented as ΔH.

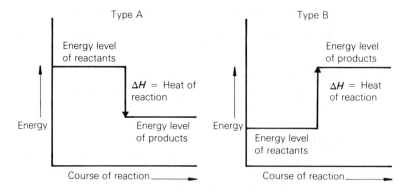

Figure 8.1   Energy changes during chemical reactions

In reactions of Type A, the products possess less energy than the reactants, and we say that the heat of reaction, ΔH, is negative. The way the reactants get rid of the excess energy is by giving out heat, and this type of reaction is called exothermic. In reactions of Type B, the products have more energy than the reactants, and we say that the heat of reaction, ΔH, is positive. Since the reactants must acquire energy in order to form the products, the reaction takes in heat: it is endothermic.

Many types of reaction involve a heat change. For each type of reaction the *heat of reaction* is defined as the heat change per mole of reactant. Heat is measured in joules (J) and kilojoules (kJ). $1000 \, J = 1 \, kJ$. *Heat of neutralisation* is the heat change that occurs when 1 mole of hydrogen ions is neutralised by a base:

$$H^+(aq) + OH^-(aq) \rightarrow H_2O \; ; \Delta H = -57.2 \, kJ \, mol^{-1}$$

Heat of combustion is the heat change that occurs when 1 mole of a substance is completely burnt in oxygen:

$$2C_2H_6(g) + 7O_2(g) \rightarrow 4CO_2(g) + 6H_2O(g)\,;\, \Delta H = -1\,560 \text{ kJ per mole of ethane}$$

Heat of formation is the heat change that occurs when 1 mole of a substance is formed from its elements:

$$S(s) + O_2(g) \rightarrow SO_2(g)\,;\, \Delta H = -297 \text{ kJ mol}^{-1}$$

Heat of solution is the heat change that occurs when 1 mole of solute is dissolved in a stated quantity of solvent.

Heat of reaction is the heat change that occurs when reaction occurs between the number of moles of reactants indicated by the equation

$$4Al(s) + 3O_2(g) \rightarrow 2Al_2O_3(s)\,;\, \Delta H = -3\,360 \text{ kJ per mole-reaction}$$

Example 1 If the heat of combustion of propane $= -2\,220$ kJ mol^{-1}, calculate the amount of heat given out when 1 000 g of propane are completely burnt.

Method:
Relative formula mass of $C_3H_8 = (3 \times 12) + (8 \times 1) = 44$

1 mole of propane has a mass of 44 g

When 44 g propane are burnt, the heat given out $= 2\,200$ kJ

When 1 g of propane burnt, the heat given out $= \dfrac{2\,200}{44}$ kJ

When 1 000 g propane are burnt, the heat given out

$$= \dfrac{1\,000 \times 2\,220}{44} \text{ kJ}$$

$$= 50\,454 \text{ kJ.}$$

Answer: The amount of heat given out is 50 000 kJ.

Example 2 The heat of combustion of butane $= -2\,877$ kJ mol^{-1}. What mass of butane must be burnt to give out 1 000 kJ?

Method:
Relative formula mass of $C_4H_{10} = (4 \times 12) + (10 \times 1) = 58$

1 mole of butane has a mass of 58 g

Since 2 877 kJ are produced by the combustion of 58 g butane,

1 kJ is produced from the combustion of $\dfrac{58}{2\ 877}$ g butane, and

1000 kJ are produced from the combustion of $\dfrac{1\ 000 \times 58}{2\ 877}$ g butane

$$= 20.16 \text{ g butane.}$$

Answer: 20.2 g butane must be burnt.

Example 3 The heat of a reaction is often measured by finding the rise in temperature produced in a known mass of water.

Heat evolved = Mass of water × Specific heat capacity × Rise in temperature

The specific heat capacity is the quantity of heat required to raise the temperature of 1 g of the substance by 1 °C. The specific heat capacity of water is 4.2 joules per gram, and the specific heat capacity of dilute solutions may be taken as the same.

When 0.29 g of propanone is burnt, the heat evolved raises the temperature of 100 g of water from 19.0 °C to 40.3 °C. What is the heat of combustion of propanone?

Method:
(a) Calculate the heat evolved.

Mass of water $= 100$ g

Specific heat capacity of water $= 4.2 \text{ J g}^{-1}$

Rise in temperature $= (40.3 - 19.0) = 21.3 \text{ °C}$

Heat evolved $=$ Mass × Specific heat capacity × Rise in temperature

$$= 100 \times 4.2 \times 21.3 = 8\ 946 \text{ J}$$

$$= 8.95 \text{ kJ}$$

(b) Calculate the molar heat of combustion.

Relative formula mass of $(CH_3)_2CO = (3 \times 12) + (6 \times 1) + 16 = 58$

1 mole of propanone has a mass of   58 g

If 0.29 g propanone give out   8.95 kJ

Then 1 g propanone gives out $\dfrac{8.95}{0.29}\,\text{kJ}$

and 58 g propanone give out $\dfrac{58\times8.95}{0.29}\,\text{kJ} = 1\,790\,\text{kJ}$

Answer: The heat of combustion of propanone is $-1\,790\text{ kJ mol}^{-1}$.

Example 4 When 100 cm³ of a 1 mol dm⁻³ solution of sodium hydroxide at 18.0 °C are added to 100 cm³ of a 1 mol dm⁻³ solution of hydrochloric acid at 18.0 °C, the temperature of the resulting salt solution is 24.8 °C. Calculate the heat of neutralisation.

Method:
(a) Calculate the heat evolved.

Volume of solution $= 200\text{ cm}^3$

Mass of solution $= 200\text{ g} = 0.2\text{ kg}$

Rise in temperature $= (24.8 - 18.0) = 6.8\ °\text{C}$

Specific heat capacity of solution $= 4.2\text{ J g}^{-1}$

Heat evolved $= \text{Mass} \times \text{Specific heat capacity} \times \text{Rise in temperature}$

$= 0.2 \times 4.2 \times 6.8 = 5.7\text{ kJ}$

(b) Calculate the heat of neutralisation.
Since 100 cm³ of 1 mol dm⁻³ NaOH(aq) contain 0.1 mole of OH⁻(aq) ions,
and 100 cm³ of 1 mol dm⁻³ HCl(aq) contain 0.1 mole of H⁺(aq) ions, the two solutions neutralise each other to produce 0.1 mole of H_2O molecules.

$$H^+(aq) + OH^-(aq) \rightarrow H_2O(l)$$

The heat given out in forming 0.1 mole of water molecules $= 5.7\text{ kJ}$

Therefore, the heat given out in forming 1 mole of water molecules $= 57\text{ kJ}$

Answer: The heat of neutralisation is -57 kJ mol^{-1}.

Example 5 This is an example of the heat change involved in a physical change of state. The amount of energy needed to melt 100 g of ice is 33.3 kJ. Calculate the heat of fusion of ice.

Method: The relative formula mass of water, H_2O, is 18.

Thus, 100 g ice are	100/18 moles of ice.
Since 100/18 moles of ice need	33.3 kJ,
1 mole of ice needs	$\dfrac{33.3 \times 18}{100}$ kJ
	= 6.0 kJ

Answer: The heat of fusion of ice is 6.0 kJ mol⁻¹.

Problems on Heat of Reaction

Section 1   No calculators or logarithm tables are needed for this section.

1. If 2.9 g of ethanol burn with the evolution of 87 kJ of heat, what is the molar heat of combustion of ethanol?

2. The molar heat of combustion of propan-1-ol is 2 010 kJ mol⁻¹. What mass of propanol must be burned to give out 402 kJ of energy?

3. The molar heat of combustion of sucrose ($C_{12}H_{22}O_{11}$) is 5 644 kJ mol⁻¹. What mass of sucrose must be burnt to give out 1 411 kJ of energy?

4. The molar heat of combustion of glucose ($C_6H_{12}O_6$) is 2 816 kJ mol⁻¹. What mass of glucose must be burnt to produce 704 kJ of energy?

5. If 13.5 g of phenyl methanol (C_7H_7OH) burn to give 505 kJ of heat, what is the molar heat of combustion of phenyl methanol?

Section 2

1. When 0.4 g of methanol is burnt, the heat evolved produces a 7 °C rise in the temperature of 304 g of water. Calculate the molar heat of combustion of methanol.

2. When 50 cm³ of 0.8 mol dm⁻³ hydrochloric acid and 50 cm³ of 0.8 mol dm⁻³ sodium hydroxide solution, both at 19 °C, are added, the temperature of the resulting solution is 24.53 °C. Calculate the molar heat of neutralisation.

3. When 2.12 g of anhydrous sodium carbonate is dissolved in 100 g of water a rise in temperature of 1.17 °C is observed. Calculate the molar heat of solution of anhydrous sodium carbonate.

4. 15.95 g of anhydrous copper(II) sulphate are dissolved in 1 dm³ of water. The temperature of the water is 18 °C, and the temperature of the solution is 19.74 °C. Calculate the molar heat of solution of copper(II) sulphate.

5. The equation for the reaction occurring when pentane is burnt is

$$C_3H_8(g) + 8O_2(g) \rightarrow 5CO_2(g) + 6H_2O(g) \; ; \; \Delta H = 3\,500 \text{ kJ}$$

Which of the following amounts of pentane, burnt completely in oxygen, will release 3 500 kJ?

(a) 1 molecule (b) 1 dm³ (c) 1 kilogram (d) 1 mole (e) 1 gram

6. 250 cm³ of 0.5 mol dm⁻³ sulphuric acid exactly neutralise 150 cm³ of a solution of sodium hydroxide. The solutions were at 18.0 °C before mixing, and, after the reaction, the temperature is found to be 26.5 °C. Calculate the molar heat of neutralisation.

Section 3   Problems from GCE O-level Papers

1. Propane, C_3H_8, is a gas at room temperature. It burns in excess air or oxygen forming carbon dioxide and water vapour and releasing 2 200 kJ/mol of propane.

   Write the equation for the reaction and calculate the energy released when 1 g of propane burns in this way.   (AEB 1977)

2. A known volume of sodium hydroxide solution was poured from a measuring cylinder into a small plastic bottle. A known volume of 2.0 M hydrochloric acid was added and the mixture was swirled. The highest temperature of the mixture was read and recorded. The results are given in the table below.

Volume of alkali	Volume of acid	Highest temperature	
cm³	cm³	°C	
5	45	22.6	graph 1
10	40	27.2	
15	35	31.8	
20	30	36.4	
25	25	34.2	graph 2
30	20	32.0	
35	15	29.8	
40	10	27.6	
45	5	25.4	

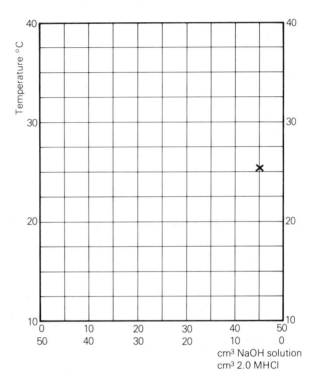

(a) Take a piece of graph paper. Mark off the temperature axis and volume axis, as shown in the grid above. Plot the results given in the table on your graph paper. The last result has been entered on the grid to help you. Draw graphs 1 and 2.

(b) By extending your graphs, find the original temperatures of the acid and alkali, say which was warmer and by how much.

(c) State the volumes of alkali and acid which will react to form a neutral solution.

(d) Calculate the concentration of the sodium hydroxide solution in grams per litre.
    (Relative atomic masses: H = 1; O = 16; Na = 23.)   (JMB)

3. (a) 100 g of an acid were analysed and found to have the following composition: 3.7 g of hydrogen, 37.8 g of phosphorus, 58.5 g of oxygen. The mass of one mole of the acid is 82 g. Use this information to find the formula of the acid.

(b) An aqueous solution of potassium hydroxide was prepared in order to investigate a reaction with this acid. The solution contained 112 g of potassium hydroxide in 1000 cm^3 of solution. Calculate the molarity of potassium hydroxide.

(c) It was found that 100 cm^3 of the potassium hydroxide solution were needed to neutralise completely 100 cm^3 of a molar solution of the acid. Show how this information can be used to write an equation for the reaction between potassium hydroxide and the acid.

(d) The temperatures of the aqueous potassium hydroxide and acid before the reaction were both 19 °C. The temperature of the 200 cm^3 of neutralised mixture reached 29 °C. Use this information to find the heat of reaction when 1 mole of the acid is completely neutralised by potassium hydroxide. (You may assume that the specific heat capacity of the solution is the same as that of water, i.e. 4.2 joules per cm^3. (London)

4. Write equations to show the complete combustion in oxygen of ethane and of ethene. Calculate the volume of oxygen needed for the complete combustion of 2000 cm^3 of ethane, all volumes being measured at room temperature and pressure.

The heats of combustion of one mole of ethane and of ethene (ethylene) are given below.

	ΔH for combustion of 1 mole
C_2H_6	-1560 kJ
C_2H_4	-1410 kJ

(a) Are these two combustion reactions exothermic or endothermic?

(b) Calculate the energy changes on complete combustion of 10 moles of ethane.

(c) Calculate the energy change on complete combustion of 140 grams of ethene. (London)

5. An important gaseous fuel is butane, C_4H_{10}. For the combustion of butane, $\Delta H = -2880$ kJ/mole.

(a) Write down the equation for butane burning in excess air, including the heat change.

(b) Calculate the quantity of heat which would be evolved when 16 litres of butane, measured at room temperature and pressure, are burnt.

(The molar volume of a gas is 24 litres at room temperature and pressure.)   (London)

6. Ammonia is manufactured from the gases nitrogen and hydrogen. The reaction may be represented by

$$N_2(g) + 3H_2(g) \rightleftharpoons 2NH_3(g) \; ; \Delta H = -93 \text{ kJ per mole-reaction}$$

(a) Will the yield of ammonia be favoured by using a high or a low temperature?

(b) What will be the energy change when 1 mole of hydrogen molecules is converted to ammonia?

(c) Is this energy evolved or absorbed?

(d) What mass of ammonia would be produced from 1 mole of hydrogen molecules?   (London)

7. The energy needed to increase the temperature of 100 g of certain metallic elements by 1 °C is given below.

Element	Relative atomic mass	Number of joules to raise 100 g by 1 °C
Magnesium	24	105
Aluminium	27	92
Copper	64	40
Molybdenum	96	25
Platinum	195	13

(a) Calculate the numbers of moles of atoms in 100 g of aluminium, copper and molybdenum.

For all FIVE metals in the table, plot a graph of the number of joules needed to raise the temperature of 100 g of metal by 1 °C against the number of moles of atoms in 100 g of metal.

(b) 100 g of the metal scandium, (Sc), were found to need 55 joules to raise its temperature by 1 °C. From your graph estimate the relative atomic mass of scandium.

If 100 g of scandium combine with 235 g of chlorine, what would you expect the formula of scandium chloride to be? Show clearly how you arrive at your result.   (London)

8. (a) Propane (C_3H_8) is a gas which is used as a fuel. If it is burnt in a plentiful supply of air,

   (i)   name the products of combustion

   (ii)   write an equation for the reaction

   (iii) state the volume of oxygen which will react completely with 4 litres of propane

   (iv)   assuming that air contains 20% oxygen by volume, what volume of air would be needed for this combustion?

   (v)   the heat of combustion is 2 200 kJ per mole of propane. Would ΔH for this reaction be given a positive or negative sign?

   (vi)   what mass of propane would need to be burnt in order to obtain 550 000 kJ?

   (b) 0.05 mole of propane occupies 1 200 cm³ at room temperature and pressure.

   (i)   What volume would it occupy if the pressure were doubled at constant temperature?

   (ii)   If the room temperature were 20 °C, at what temperature would the volume be doubled at constant pressure? (London)

9. It is possible to titrate an acid with an alkali by measuring the temperature changes that occur when they are mixed in different proportions. In an experiment of this type, 5.0 cm³ of 2 M aqueous sodium hydroxide and 45.0 cm³ of hydrochloric acid of unknown concentration were measured out into separate beakers. The two solutions were then mixed and the temperature rose by 2.0 °C. The experiment was repeated several times, using different volumes of acid and alkali but keeping the total volume 50.0 cm³ in each case.

The highest temperature rise was found to be 10.0 °C, and occurred with 20.0 cm³ of sodium hydroxide and 30.0 cm³ of hydrochloric acid.

(a) Which graph of the results would be obtained?

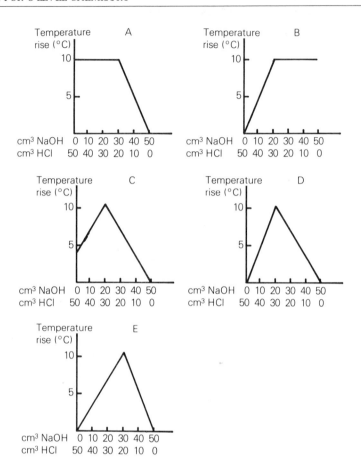

(b) The reaction may be represented by the equation

$$NaOH + HCl \rightarrow NaCl + H_2O$$

What was the concentration of the hydrochloric acid?

(i) 0.75 M   (ii) 1.00 M   (iii) 1.33 M   (iv) 2.00 M   (v) 3.00 M

(c) If the hydrochloric acid had been of exactly the same concentration (in moles per litre) as the sodium hydroxide, which mixture would have given the maximum temperature rise?

	Volume of NaOH (cm³)	Volume of HCl (cm³)
(i)	10.0	40.0
(ii)	20.0	30.0
(iii)	25.0	25.0
(iv)	30.0	20.0
(v)	40.0	10.0

(London)

9. The Rates of Chemical Reactions

Reaction Rates

Steps can be taken to speed up a chemical reaction, or to slow it down. There are four factors which influence the rate of a chemical reaction.

1 The particle size of the solid reactants

Marble chips react with hydrochloric acid, according to the reaction

$$CaCO_3(s) + 2HCl(aq) \rightarrow CaCl_2(aq) + CO_2(g) + H_2O(l)$$

Small chips react faster than large chips, and powder reacts faster still. This is because small lumps have a greater surface area per gram of marble than large lumps. There is more surface for the acid to attack, and small lumps therefore react faster than big lumps.

2 The concentrations of the reacting solutions

When substances react in solution, the rate of reaction is proportional to the concentration of each reactant. This can be shown for a reaction such as the formation of hydrogen in the reaction between magnesium and a dilute acid:

$$Mg(s) + H_2SO_4(aq) \rightarrow H_2(g) + MgSO_4(aq)$$

The rate at which hydrogen can be collected is doubled if the concentration of sulphuric acid is doubled.

3 Temperature

Reactions take place faster if the temperature is increased. Reactions which take 5–10 minutes for completion will take place about twice as fast if the temperature is raised by 10 °C.

4 Catalysts

A catalyst is a substance which speeds up a chemical reaction. For example, manganese(IV) oxide speeds up the dissociation of hydrogen peroxide to give oxygen. Although it plays a part in the reaction, a catalyst is not used up, and can be recovered unchanged at the end of the reaction

$$2H_2O_2(aq) \rightarrow 2H_2O(l) + O_2(g)$$

105

In most catalysed reactions, a very small quantity of catalyst is all that is required. The reaction does not go faster still if you add more catalyst.

Graphs which show the extent of reaction plotted against time

The rate of reaction of a substance in solution is proportional to the concentration of the solution. As more and more of the reactant is used up, the reaction becomes slower and slower. The progress of the reaction has the form shown in Figure 9.1.

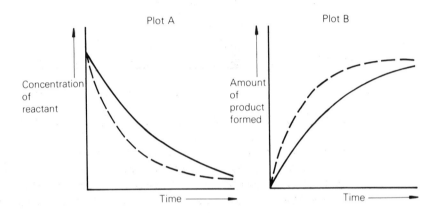

Figure 9.1   Graphs of extent of reaction against time

Plot A shows the way in which the concentration of a reactant decreases as the reaction takes place. Plot B shows how the amount of product increases as the time interval since the beginning of the reaction increases. For example, if you collect the oxygen formed in the decomposition of hydrogen peroxide in a syringe, and read off its volume at various intervals of time after the beginning of the reaction, the plot of the volume of gas formed against time will have the form shown in Plot B.

If the reaction is speeded up, the reactant will be used up in a shorter time (as shown by the broken line in A), and the product will be formed in a shorter time (as shown by the broken line in B). Methods of speeding up the reaction do not affect the amount of product formed. 34 g of hydrogen peroxide will give a maximum of 11.2 dm^3 of oxygen (at s.t.p.) whether the reaction occurs at room temperature or at 80 °C, whether the reaction occurs slowly over a period of weeks or quickly in the presence of a catalyst. You cannot create more product by altering the conditions.

Example 1   Figure 9.2 shows a plot of the volume of oxygen produced from the decomposition of hydrogen peroxide against the time for which the reaction has been occurring. State whether A or B or C represents the fastest reaction.

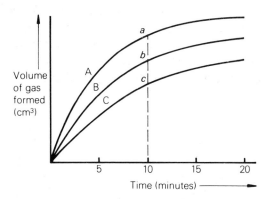

Figure 9.2   Plot of volume of oxygen against time

Method: Read upwards from a time of 10 minutes on the time axis. In reaction A, a cm³ of gas have been formed; in reaction B, b cm³ of gas have been formed and in reaction C, c cm³ of gas have been formed. You can see that volume a is more than volume b, which is more than c, so that reaction A is faster than B, which is faster than C.

Example 2

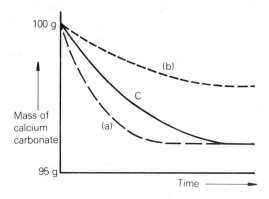

Figure 9.3   Plot of mass of calcium carbonate against time

Curve C in Figure 9.3 shows how the mass of calcium carbonate decreases with time when 100 g of marble chips are allowed to react with 50 cm³ of 2 mol dm⁻³ hydrochloric acid. On the graph, mark the plots you would obtain (a) by carrying out the reaction at 30 °C, (b) by using 50 cm³ of 1 mol dm⁻³ hydrochloric acid.

Method: At a higher temperature, the rate of reaction will be faster. Curve (a), therefore, lies to the left of curve C.

Using a lower concentration of acid, the rate will be slower, and plot (b), therefore, lies to the right of curve C. With 50 cm³ of 1 mol dm⁻³ acid (instead of 2 mol dm⁻³) only half as much calcium carbonate will be able to react, and there will be only half the decrease in mass. Curve (b) levels off after a decrease of 2.5 g, instead of 5 g.

Example 3   The volume of carbon dioxide formed by the reaction of an excess of calcium carbonate with 50 cm³ of 0.1 mol dm⁻³ hydrochloric acid is plotted against time, in Figure 9.4.

(a) What volume of gas has been formed after 5 minutes?

(b) How long does the reaction take to produce 50 cm³ of gas?

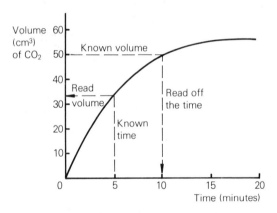

Figure 9.4   Plot of volume of carbon dioxide against time

Method: (a) Draw a vertical line from the stated time of 5 minutes. Where it cuts the graph, draw a horizontal line across to the volume axis, and read off the volume.

Answer: 32 cm³ of carbon dioxide.

(b) Draw a horizontal line across from the stated volume, 50 cm³, to cut the graph. Where it cuts the graph, drop a vertical line down to the time axis, and read off the time.

Answer: 10 minutes.

Problems on Rate of Reaction

Questions from GCE O-level Papers

1. Two experiments were carried out, using in each case the same volume of a solution of hydrogen peroxide.

   Experiment A: The solution was made alkaline at room temperature.

   Experiment B: A small amount of manganese(IV) oxide was added at room temperature.

   The volume of oxygen evolved, measured at room temperature and pressure, was plotted against time for each experiment. The graphs obtained were as shown in the following diagram:

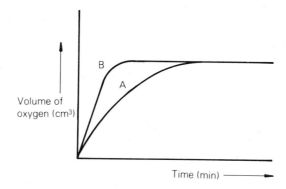

(a) In which experiment, A or B, did the reaction begin more rapidly? Using the graphs, explain how you arrived at your answer.

(b) State and explain the function of the manganese(IV) oxide in experiment B.

(c) Both experiments eventually gave the same volume of oxygen. Using the following equation for the decomposition of hydrogen peroxide:

$$2H_2O_2(aq) \rightarrow 2H_2O(l) + O_2(g)$$

calculate the volume of oxygen evolved at room temperature and pressure by the complete decomposition of 1 000 cm³ of hydrogen peroxide solution containing 0.68 g/l. (A_r(H) = 1; A_r(O) = 16. Molar volume of a gas at room temperature and pressure is 24 litres.)   (Welsh JEC)

2. Experiments were carried out in which portions of powdered calcium carbonate were reacted with hydrochloric acid and the volume of carbon dioxide evolved was measured at intervals in a graduated container at room temperature and pressure. The relevant data and the graphical representation of the results are given below.

Experiment	Mass of calcium carbonate	Volume and concentration of hydrochloric acid
A	1.0 g	40 cm³ of 1.0 mol/l
B	1.0 g	20 cm³ of 2.0 mol/l
C	0.5 g	20 cm³ of 0.5 mol/l

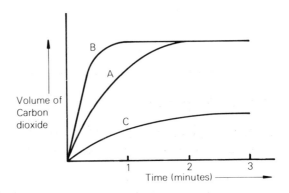

(a) In which experiment does the reaction begin most rapidly? From the graph, explain how you arrived at your answer.

(b) Give the reason for the differences in the initial rates of reaction in the three experiments.

(c) Write the equation for the reaction occurring.

(d) What fraction of a mole of (i) calcium carbonate, (ii) hydrochloric acid was used in each of the experiments, A, B and C?

(e) From your answers to (c) and (d):
   (i) deduce, giving your reasons, in which experiments there was sufficient acid to react with all the calcium carbonate
   (ii) calculate the volume of carbon dioxide obtained in experiment A after the reaction has ceased. (Molar volume of a gas at room temperature and pressure is 24 litres.)

(Welsh JEC)

3. This question is about a set of experiments in which magnesium was added to 2 molar hydrochloric acid. A fixed mass of magnesium (0.6 g) was used in each experiment, but a different volume of acid was chosen each time. A gas was formed and its volume was measured at room temperature and pressure. The results of the experiments are given in the table below.

Volume of 2 M HCl used, in cm³	Volume of gas formed, in cm³
5	120
15	360
25	600
35	600
45	600

(a) Name the gas formed.

(b) Plot a graph of volume of gas formed against volume of acid used. Volume of gas should be plotted along the vertical axis.

(c) Use your graph to predict
   (i) the volume of gas produced if 50 cm³ of 2 M hydrochloric acid is added to 0.6 g of magnesium
   (ii) the volume of 2 M hydrochloric acid which must be added to 0.6 g of magnesium to produce 480 cm³ of gas.

(d) Use your graph to find the volume of 2 M hydrochloric acid which would just dissolve 0.6 g of magnesium. You must explain clearly how you have used the graph.

(e) Use the experimental results to predict
  (i) the volume of 2 M hydrochloric acid which would just dissolve one mole of magnesium atoms
  (ii) the volume of gas which would be formed when 1 mole of magnesium atoms dissolves in hydrochloric acid.

(f) Use your answer in part (e) to write an equation for the reaction between magnesium and hydrochloric acid. You *must* show clearly how you have used the information from part (e). (London)

4. Some coarse lumps of manganese(IV) oxide were added to 20 cm³ of a dilute aqueous solution of hydrogen peroxide in a suitable apparatus at room temperature and pressure. Oxygen was set free in accordance with the equation

$$2H_2O_2(aq) \rightarrow 2H_2O(l) + O_2(g)$$

The volume was noted every minute, and a graph of the volume of oxygen produced against time from the start of the experiment was plotted as shown below.

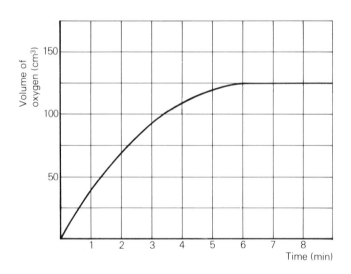

(a) What volume of oxygen was finally produced?

(b) How many moles of O_2 molecules does this represent?

(c) What mass of hydrogen peroxide was present in the original solution?

(d) Find the time taken for half of the hydrogen peroxide to be decomposed. State briefly how you arrived at your answer.

(e) Without altering the mass of manganese(IV) oxide or the mass of hydrogen peroxide, state TWO changes you could make to increase the rate of decomposition.

(f) Copy the graph onto a piece of graph paper, and sketch in the curve that you might expect as a result of either of the changes that you suggest in (e).   (London)

5. 10 cm^3 (an excess) of 1 M HCl were added to 0.048 g of magnesium ribbon in a suitable apparatus at 20 °C and atmospheric pressure. The volume of hydrogen formed was noted every 30 seconds, and a graph of volume of gas produced against time from the start of the experiment was plotted as shown below.

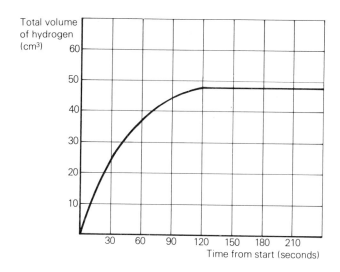

(a) At what time was the action most rapid?

(b) How many seconds had elapsed before all the magnesium had dissolved?

(c) What volume of hydrogen had been produced when all the magnesium had dissolved?

(d) How many seconds had elapsed before half of the magnesium had dissolved?

(e) How many moles of atoms are there in 0.048 g of magnesium?

(f) An equation for the reaction is

$$Mg(s) + 2HCl(aq) \rightarrow MgCl_2(aq) + H_2(g)$$

What volume of 1 M HCl would be required to react completely with 0.048 g of magnesium?

(g) Copy the graph onto a piece of graph paper, and sketch in the curve you would expect if the experiment was repeated in an identical manner, except that the temperature was 30 °C.

(London)

6. A conical flask containing excess dilute nitric acid is weighed after a lump of magnesium carbonate has been added. A loose plug of glass wool is placed in the mouth of the flask to prevent the loss of acid spray and the mass of the apparatus is recorded at regular intervals. The graph of the 'mass of the flask + contents' against 'time' is plotted. Which one of the following curves A, B, C, D or E is most likely to be the graph obtained?

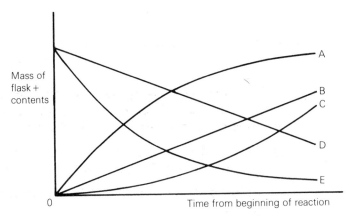

(Cambridge)

7. The decomposition of hydrogen peroxide is catalysed by manganese(IV) oxide (MnO_2).

$$2H_2O_2 \rightarrow 2H_2O + O_2$$

Three separate experiments were performed in which 1.0 g samples of powdered MnO_2 were added to mixtures of hydrogen peroxide solution and water. The total volume of solution was 100 cm³ in each case. The volume of oxygen produced during the first 10 seconds after adding the MnO_2 to each solution was measured at constant room temperature and pressure.

(a) The results of the experiments are shown in the following table:

Volume of H_2O_2 solution taken (cm³)	Volume of water taken (cm³)	Volume of oxygen produced in the first 10 seconds (cm³)
100	0	120
50	50	60
25	75	30

The volume of oxygen produced in 10 seconds is a measure of the rate of the reaction. If the volume of H_2O_2 solution is x, the rate of reaction is proportional to:

(i) $\dfrac{1}{x}$ (ii) $\dfrac{1}{x^2}$ (iii) $\sqrt{x}$

(iv) x (v) x^2

(b) If the experiment were repeated, which change would NOT affect the volumes of oxygen produced (measured at the same temperature and pressure) in the first 10 seconds?

(i) raising the temperature of the hydrogen peroxide solutions before adding the manganese(IV) oxide

(ii) adding 1.0 g of manganese(IV) oxide in lumps instead of powder

(iii) using twice as much hydrogen peroxide and making up to 200 cm³ where necessary by adding water

(iv) using 1.0 g of some other catalyst

(v) using a more concentrated solution of hydrogen peroxide to make up the mixtures.

(c) If different masses of powdered manganese(IV) oxide were added to 100 cm³ samples of hydrogen peroxide solution which were all of the same concentration, which graph of the results would you expect to obtain?

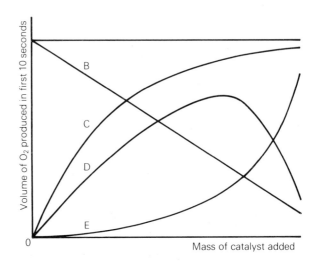

(d) Which fact provides evidence that the MnO_2 is acting as a catalyst?

 (i)   The mass of MnO_2 does not change during each experiment.
 (ii) The volume of oxygen produced varies with the concentration of the hydrogen peroxide.
 (iii) The hydrogen peroxide becomes more dilute as the reaction proceeds.
 (iv) The volume of oxygen produced is greater than could be obtained by decomposing the MnO_2.
 (v) MnO_2 does not give off oxygen unless heated to a high temperature.

(e) In another experiment, the total volume of oxygen produced from 100 cm³ of H_2O_2 solution was measured. This was found to be 500 cm³. Hydrogen peroxide decomposes in the following way:

$$2H_2O_2 \rightarrow 2H_2O + O_2$$

(H = 1, O = 16; 1 mole of oxygen molecules occupies $24\,000$ cm³ at room temperature and pressure.)

The number of grams of hydrogen peroxide present in the 100 cm³ sample of solution was

 (i) $34 \times 500 \times 12\,000$

 (ii) $34 \times \dfrac{500}{12\,000}$

(iii) $34 \times \dfrac{12\,000}{500}$

(iv) $34 \times \dfrac{24\,000}{500}$

(v) $34 \times \dfrac{500}{24\,000}$

(London)

8.

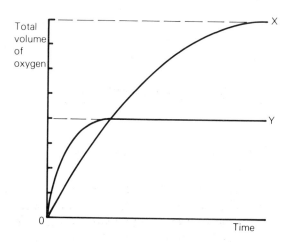

Graphs X and Y represent the results of two experiments demonstrating the catalytic decomposition of hydrogen peroxide. Assuming that *all other* conditions are kept constant, which one of the following is a correct explanation of the different results?

Experiment X	*Experiment* Y
(a) 50 cm³ of 1.0 mol/dm³ hydrogen peroxide were used.	25 cm³ of 2.0 mol/dm³ hydrogen peroxide were used.
(b) 1.0 g of manganese(IV) oxide was used.	0.50 g of manganese(IV) oxide was used.
(c) The reaction was carried out at 60 °C.	The reaction was carried out at 30 °C.
(d) 50 cm³ of 1.0 mol/dm³ hydrogen peroxide were used.	12.5 cm³ of 2.0 mol/dm³ hydrogen peroxide were used.
(e) The catalyst was in lumps.	The catalyst was finely divided.

(Cambridge)

9. In an investigation to show how the surface area of marble chips affects the rate of its reaction with hydrochloric acid, the apparatus is set up as shown in the figure. The mass of the flask and its contents is observed at half-minute intervals after the addition of the acid until the mass becomes constant. The experiment is repeated using smaller marble chips, and a third experiment is performed with powdered marble, using the same mass each time. In each experiment, 50 cm³ of 2 M hydrochloric acid was added, and marble remained at the end.

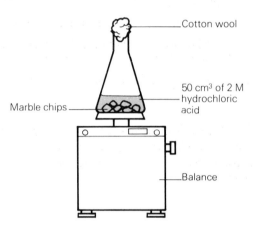

Results

Time (min)	1st experiment (large chips)	2nd experiment (small chips)	3rd experiment (powder)
0	0	0	0
0.5	0.45	0.64	0.80
1.0	0.75	0.95	1.20
1.5	0.96	1.23	1.52
2.0	1.15	1.44	1.76
2.5	1.32	1.60	1.88
3.0	1.46	1.76	1.96
3.5	1.60	1.88	2.00
4.0	1.70	1.96	2.00
4.5	1.80	2.00	2.00
5.0	1.88	2.00	2.00
5.5	1.92	2.00	2.00
6.0	1.97	2.00	2.00
7.0	1.99	2.00	2.00
8.0	2.00	2.00	2.00

(a) Using the figures in the table of results, plot graphs for all three experiments on the same piece of graph paper.

(b) For each experiment, state the time required for half the acid to react.

(c) Draw tangents to each graph at these times. Work out the rates at which the mass is decreasing at these instants.

(d) Which of the samples of marble reacts at the highest rate and why?

(e) Why is there a decrease in mass during these experiments?

(f) In all three experiments, the final loss in mass is the same. Explain why there is no further loss.

(g) Why is a plug of cotton wool placed in the neck of the flask?

(Cambridge)

10. Calcium carbonate was placed in a flask on a top-pan balance, and dilute hydrochloric acid was added. The total mass of the flask and its contents was recorded every five seconds. The figure shows a plot of the results.

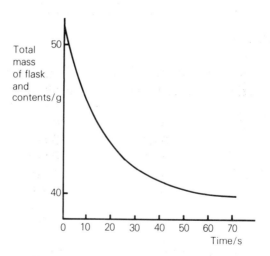

At which one of the following times was the reaction fastest?

(a) 10 s   (b) 20 s   (c) 30 s   (d) 40 s   (e) 50 s.

(Cambridge)

11.   The graph represents the progress of reaction between calcium carbonate and an excess of hydrochloric acid. The curve shows how the total volume of carbon dioxide liberated (at s.t.p.) varied with time.

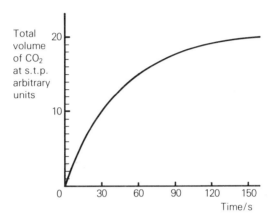

How long did it take for half of the calcium carbonate to react?

(a) 30 s   (b) 60 s   (c) 90 s   (d) 120 s   (e) 150 s

(Cambridge)

10. Solubility

Solubility

Many salts dissolve in water, to form a solution. As more and more salt is added, a point is reached where the solution will dissolve no more salt. It is then said to be a saturated solution. A *saturated solution* is one which contains the maximum amount of solute that can be dissolved at that temperature. The best way to make sure that a solution is saturated is to have an excess of solute present. For most salts, the solubility increases with temperature.

Solubility is defined as the number of grams of solute required to saturate 100 g of solvent at a stated temperature. Saturated solutions can be prepared at various temperatures and analysed to find the concentration of dissolved solute. The measured solubilities can be plotted against temperature on a graph. The graph is called a solubility curve. Figure 10.1 shows solubility curves for a number of substances over temperatures of 0 °C to 100 °C.

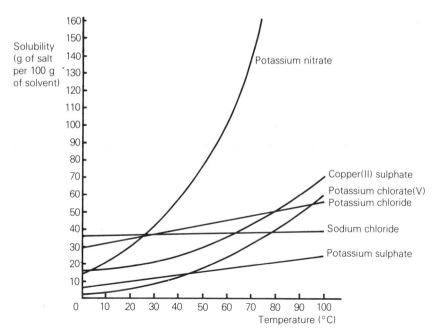

Figure 10.1   Solubilities of some salts in water

Example 1   If 8 g of potassium chloride saturate 20 g of water at 40 °C, what is the solubility of potassium chloride?

Method: Problems of this kind are a simple ratio type of calculation.

If 20 g water are saturated by 8 g potassium chloride,

$$100 \text{ g water are saturated by } \frac{100 \times 8}{20} = 40 \text{ g potassium chloride.}$$

Answer: Solubility of potassium chloride at 40 °C = 40 g per 100 g of water.

Example 2   100 g of water at 80 °C are saturated with potassium nitrate. What mass of potassium nitrate will crystallise out if the solution is cooled to 20 °C?

Method: Look at the solubility curves in Figure 10.1. Draw a vertical line from 80 °C to cut the solubility curve of potassium nitrate. On the solubility axis, read off what mass of potassium nitrate is dissolved. You can see that 170 g of potassium nitrate are dissolved in 100 g of water at 80 °C. Draw a vertical line from 20 °C to cut the solubility curve. You can read off a value of 35 g per 100 g of water. On cooling from 80 °C to 20 °C, the solubility drops from 170 to 35 g per 100 g of water. The difference, 135 g of potassium nitrate, comes out of solution.

Answer: 135 g of potassium nitrate come out of solution on cooling 100 g of water saturated with this salt from 80 °C to 20 °C.

Example 3   The following table gives the solubility of hydrogen sulphide in g per dm^3 of water at a number of temperatures:

Temperature (°C)	0	20	40	60	70	80	90
Solubility (g dm⁻³)		7.07	3.85	2.36	1.48	1.10	0.765 0.41

(a) Plot a solubility curve for hydrogen sulphide in water at different temperatures.

(b) From the curve, read off the solubility of hydrogen sulphide at 30 °C and at 50 °C.

(c) Calculate the concentrations of the two saturated solutions.

Method: Plot solubility along the vertical axis. Let 1 cm represent 1 g per dm³. Plot temperature along the horizontal axis. Let 1 cm represent 10 °C. You will obtain the graph shown in Figure 10.2.

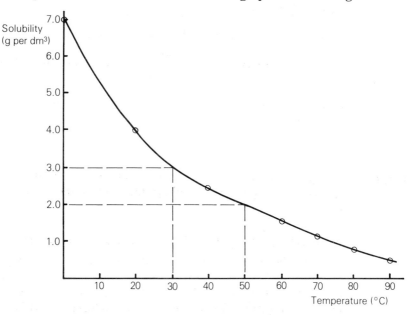

Figure 10.2   Solubility curve for hydrogen sulphide

(a) Draw a vertical line at 30 °C, and find where it cuts the graph. Read off the solubility. Answer = 3.0 g dm⁻³.

(b) Draw a vertical line at 50 °C, and find where it cuts the graph. Read off the solubility. Answer = 2.0 g dm⁻³.

(c) Relative formula mass of H_2S = (2 + 32) = 34.

Concentration of solution (a) (mol dm⁻³)
$$= \frac{\text{Mass of solute per dm}^3}{\text{Relative formula mass}} = \frac{3}{34} = 0.088 \text{ mol dm}^{-3}$$

Concentration of solution (b) (mol dm⁻³)
$$= \frac{\text{Mass of solute per dm}^3}{\text{Relative formula mass}} = \frac{2}{34} = 0.059 \text{ mol dm}^{-3}$$

Problems on Solubility

Questions from GCE O-level Papers

1. The table gives the solubility of potassium chlorate, in g per 100 g of water, at various temperatures.

Temperature (°C)	10	20	30	40	50	60	70	80	90
Solubility g per 100 g of water	5	7.5	10.5	14	18.5	24	30	38	46

On graph paper, construct a solubility curve for potassium chlorate. Use your graph to answer the following questions ON THE GRAPH PAPER:

(a) What is the solubility of potassium chlorate at 0 °C?

(b) A solution containing 35 g of potassium chlorate dissolved in 100 g of water is cooled. At what temperature will crystals start to form?

(c) The same solution is cooled to 30 °C. What mass of crystals will be deposited? (Southern UJB)

2. (a) Using graph paper and on *the same axes,* draw and label solubility curves for potassium chloride and potassium chlorate(V) from the following data:

Temperature (°C)	0	20	40	60	80	100
Potassium chloride g/100 g water	28.0	33.0	38.0	43.0	48.0	53.0
Potassium chlorate(V) g/100 g water	4.0	7.5	14.0	25.0	41.0	59.0

(b) You are provided with 300 g of water and a solid mixture of 28.0 g of potassium chlorate(V) with 84.0 g of potassium chloride. Describe how you would obtain a sample of potassium chlorate(V) from the mixture. With the aid of your solubility curve calculate the maximum mass of potassium chlorate(V) you would expect to obtain if the solution was cooled to 0 °C. (AEB 1978)

3. (a) The table below gives information about the solubility of sulphur dioxide in water at different temperatures. The masses of sulphur dioxide given in the table are those required to produce a saturated solution.

Temperature (°C)	10	15	25	30	40	50	60
Solubility of sulphur dioxide in grammes per litre of solution	154	126	90	77	58	42	32

Plot a graph of the solubility of sulphur dioxide against temperature, with the temperature along the x axis (the horizontal axis). Use your graph to estimate

(i) the mass of sulphur dioxide which would dissolve in 1 litre of saturated solution at 45 °C

(ii) the temperature at which a saturated solution of sulphur dioxide would contain 64 g of sulphur dioxide in 1 litre of solution. What is the molarity of this solution?

(b) On cooling a solution of sulphur dioxide in water, crystals separate out which are found to contain 37.2% by mass of sulphur dioxide and 62.8% by mass of water.

It is thought that these crystals can be represented by a formula of the type $SO_2 \cdot xH_2O$. Calculate the value of x in this formula.

(London)

4. A solid acid can be represented by the formula H_2X, and it has a formula weight of 130. At 25 °C, 10 g of the acid would dissolve in 100 g of water. At 85 °C, 30 g of the acid would dissolve in 100 g of water. The following experiments were carried out to determine the solubilities at these two temperatures.

I: 25 g of a saturated solution of the acid in water at 70 °C were evaporated to dryness. The residue of acid had a mass of 5 g.

II: 25 g of a saturated solution of the acid in water at 40 °C needed 50 cm³ of molar (1 M) sodium hydroxide for complete neutralisation. The equation for this neutralisation reaction is

$$H_2X + 2NaOH \rightarrow Na_2X + 2H_2O$$

(a) Determine the mass of acid which would dissolve in 100 g of water at 70 °C and at 40 °C.

(b) Plot a graph of the number of grams of acid dissolving in 100 g of water against the temperature. Use a scale of 2 cm for 10 °C along the x axis (horizontal axis) and 2 cm for 5 g along the y axis.

(c) Use your graph to predict:

(i) the maximum mass of acid which would dissolve in 100 g of water at 0 °C.

(ii) the mass of acid which would separate out if 100 g of water saturated with acid at 55 °C were cooled to 10 °C.

(London)

5. The table below gives information about the solubility of the gas, hydrogen chloride, in water at different temperatures.

Temperature (°C)	0	10	30	40	50	60
Mass in grams dissolving in 1 litre of water	825	770	675	635	595	560

(a) Plot a solubility curve for hydrogen chloride in water at different temperatures. Make solubility the vertical axis, and start the scale at 500 g/litre.

From your graph, determine the number of grams of hydrogen chloride which will dissolve in 1 litre of water at 20 °C, and calculate how many moles of hydrogen chloride this represents.

(b) If it is assumed that there is no change in volume when the hydrogen chloride dissolves in water, what is the molarity of the solution at 20°C?

There is in fact a volume change when the hydrogen chloride dissolves in water. What extra information would you need to enable you to calculate the molarity of the solution *without making the assumption above?*

How would you attempt to determine the molarity of the solution directly in order to verify your result experimentally?

(London)

6. This question concerns an investigation of the solubility of various salts in water at various temperatures.

The *solubility* of a substance in water at a particular temperature is the maximum number of grams of the substance which can be dissolved in 100 g of water at that temperature in the presence of excess of the solid substance.

The solubility of potassium chloride at 30 °C was first determined. Some water was maintained at 30 °C in a beaker and potassium chloride was added until no more would dissolve. Some of the solution was carefully poured off from the excess solid into a weighed evaporating basin. The solution in the evaporating basin was evaporated to dryness over a Bunsen flame. The mass of the evaporating basin and solid potassium chloride remaining at the end of the experiment was determined. From the results, the solubility of potassium chloride at 30 °C was determined. The experiment was then repeated at other temperatures.

The whole experiment was repeated with other salts.

(a) In one experiment 16.0 g of a SATURATED solution at a particular temperature was found to contain 4.0 g of solid. The number of grams of the solid which would dissolve in 100 g of water would be:

(i)   12.0   (ii) 20.0   (iii) 25.0
(iv) 33.3   (v) 66.7

(b) In determining the solubility of potassium chloride at 30 °C as described above, one important measurement was omitted. This was to measure the:
   (i)   mass of water originally in the beaker at 30 °C
   (ii) total mass of potassium chloride originally added to the beaker
   (iii) temperature of the room
   (iv) mass of the solution and evaporating basin before evaporation to dryness
   (v) temperature of the boiling solution in the evaporating basin

The results of the experiments are shown in the graphs below. Use the graphs to answer questions (c)–(e).

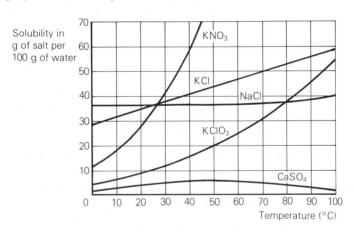

(c) What is the maximum number of grams of $KClO_3$ which will dissolve in 50 g of water at 50 °C?
   (i)   5   (ii) 10   (iii) 20
   (iv) 45   (v) 90

(d) The most soluble salt at 20 °C is:
   (i) $CaSO_4$ (ii) $KClO_3$ (iii) NaCl
   (iv) KCl   (v) KNO_3

(e) If 50 g of KCl were dissolved in 100 g of water at 100 °C and the solution cooled to 40 °C, the number of grams of solid deposited would be:
   (i)   10   (ii) 15   (iii) 25
   (iv) 30   (v) 50   (London)

7. The graph shows the solubility of ammonia in water at different temperatures. (The masses of ammonia are those required to produce a saturated solution in 1 dm³ of solution.)

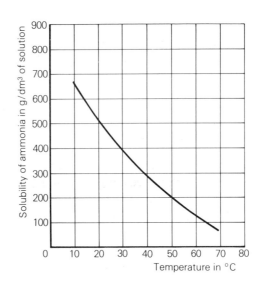

(a) What do you understand by the term *saturated solution* for a gas?

(b) Compare the effect of temperature on the solubility of a gas in water with its effect on the solubility of a solid (such as potassium nitrate) in water.

(c) What is the concentration, in mol/dm³, of a saturated solution of ammonia in water at room temperature (20 °C)?

(d) Estimate from the graph the solubility of ammonia in water at 5 °C.

(e) When a concentrated solution of ammonia in water is cooled, crystals of hydrated ammonia are formed containing 48.6% of ammonia by mass. What is the formula of these crystals?

(Cambridge)

11. Radioactivity

Radioactivity

A large number of elements are *radioactive*: their nuclei are unstable and split to form two new nuclei. This type of reaction is called a *nuclear reaction.* It is quite different from a chemical reaction in which the atoms stay the same and only the bonding changes. Sometimes, protons, neutrons and electrons fly out when the original nucleus divides. The particles and energy given out are called *radioactivity,* and the breaking-up process is called *radioactive decay.*

There are three types of radiation: α-rays, β-rays and γ-rays. β-rays and γ-rays are penetrating rays, and can be detected by means of an instrument called a Geiger-Müller counter. Each time a β-ray or a γ-ray is emitted, it ionises the gas inside the Geiger-Müller counter, and a pulse of electricity flows through the counter. The pulses of electricity can be fed into a loudspeaker to produce a click or counted on an electronic counter.

The results of measurements of radioactive decay always have the form shown in Figure 11.1.

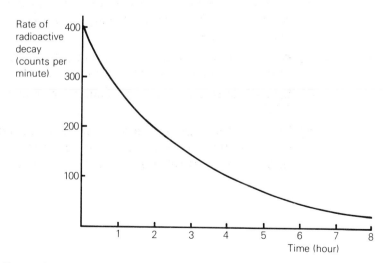

Figure 11.1 A decay curve for a radioactive isotope

You can see that

the time taken for the counts to fall from 400 to 200 c.p.m. is 2 hours, and

the time taken for the counts to fall from 200 to 100 c.p.m. is 2 hours, and

the time taken for the counts to fall from 100 to 50 c.p.m. is also 2 hours.

The time taken for the rate of radioactive decay to fall to half its value is the same, no matter what the original rate of decay is. This time is called the *half-life*. Radioactive elements differ enormously in their half-lives, from a fraction of a second to millions of years.

You often see elements referred to as 'radioactive isotopes'. Some elements have atoms of more than one kind. The different kinds of atoms are called *isotopes*. They differ in the number of neutrons in the nucleus. Thus, they have the same atomic number (the number of protons) and different mass numbers (the number of protons plus the number of neutrons). They are represented as

$$\substack{\text{Mass number} \\ \text{Atomic number}} \text{Symbol}$$

For example, the isotopes of carbon are $^{12}_{6}C$, $^{13}_{6}C$ and $^{14}_{6}C$. They are often referred to as carbon-12, carbon-13 and carbon-14.

Examples of Problems on Radioactivity: Type 1

Problems on radioactivity are very simple. You may be told the half-life of a radioactive isotope and asked to work out how the count rate will decrease with time. You may, given the half-life, be asked to calculate how the mass of a radioactive isotope decreases with time. Another kind of problem is to work out the half-life of a radioactive decay from a plot of counts per minute against time.

Example 1   The half-life of bromine-82 is 36 hours. A sample of potassium bromide solution containing bromine-82 is put into a Geiger-Müller counter. The count rate is 160 c.p.m. What will be the value of the count rate 72 hours later?

Method:
Count rate = 160 c.p.m.
Half-life   =   36 hours.
The count rate will have fallen to half its value in 36 hours.
Count rate after 36 hours = 80 c.p.m.

In a further 36 hours, the count rate will fall to half its value, i.e. from 80 to 40 c.p.m.
Count rate after 72 hours = 40 c.p.m.

Answer: Count rate after 72 hours = 40 c.p.m.

Example 2   Caesium-137 has a half-life of 30 years. A sample of caesium chloride contains 2 g of caesium-137. What will be the mass of caesium-137 left after 90 years?

Method:
Mass of radioactive isotope = 2 g
Half-life   = 30 years
Time span   = 90 years = 3 half-lives
Mass of radioactive isotope left after 3 half-lives = $2 \times \frac{1}{2} \times \frac{1}{2} \times \frac{1}{2}$
                                                                                    = 0.25 g.

Answer: 0.25 g of caesium-137 will remain after 90 years.

Example 3   A solution of radioisotope was placed in a Geiger-Müller counter, and the count rate was measured at intervals over a number of days. The results are shown in Table 11.1. Choosing a suitable scale, plot the rate of radioactive decay against time. From your graph, find the half-life of the nuclear reaction.

Table 11.1

Time (days)	Decay rate (c.p.m.)
0	1000
1	820
2	660
3	520
4	420
5	340
6	260
8	170

Method: The first thing you have to do is to choose a scale. Along the x axis, 1 cm = 1 day will spread the results out nicely. Along the y axis, 1 cm = 100 c.p.m. will be suitable.

The plot of the results is shown in Figure 11.2.

At zero time, the decay rate is 1000 c.p.m. The half-life of the radioactive decay is the time taken for the c.p.m. to fall to half this value, i.e. to 500 c.p.m. Draw a horizontal line from 500 c.p.m. across to intersect the graph. Drop a vertical line to the x axis to find the time which has passed. The time is 3.2 days.

Now find the time taken for the count rate to fall from 500 to 250 c.p.m. Draw a horizontal line across to cut the graph. Drop a vertical line to the x axis. Read off the time. The time is 6.4 days. To fall from 500 c.p.m. to 250 c.p.m. takes from 6.4 to 3.2 days, that is 3.2 days.

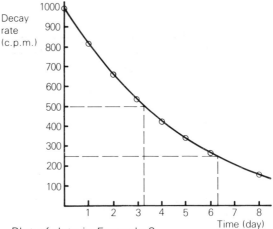

Figure 11.2   Plot of data in Example 3

The graph will tell you the time taken for the count rate to fall from 800 c.p.m. to 400 c.p.m., from 600 to 300 c.p.m. and so on. Whichever part of the graph you use will give the same answer for the half-life.

Examples of Problems on Radioactivity: Type 2

There is another type of numerical problem on radioactivity, in which you are asked to fill in the missing mass numbers and atomic numbers in the equation for a nuclear reaction.

Example 1   The radioactive isotope of sodium decays according to the equation

$$^{24}_{11}\text{Na} \rightarrow {}^{0}_{-1}\text{e} + {}^{m}_{n}X$$

Fill in the mass number, m, and the atomic number, n, of the element, X.

Method: The sum of the atomic numbers on the left-hand side of the equation equals the sum of the atomic numbers on the right-hand side of the equation. Therefore,

$$24 = 0 + m$$
$$m = 24$$

Sum of mass numbers on LHS = Sum of mass numbers on RHS. Therefore,

$$11 = -1 + n$$
$$n = 12$$

Answer: The isotope formed is $^{24}_{12}X$.

Example 2 Uranium-238 decays according to the equation below. Supply the mass number and atomic number of the thorium isotope formed.

$$^{238}_{92}U \rightarrow {}^{4}_{2}He + {}^{a}_{b}Th$$

Method:
Sum of atomic numbers on LHS = Sum of atomic numbers on RHS. Therefore,

$$238 = 4 + a$$
$$a = 234$$

Sum of mass numbers on LHS = Sum of mass numbers on RHS. Therefore,

$$92 = 2 + b$$
$$b = 90$$

Answer: The isotope formed is $^{234}_{90}Th$.

Problems on Radioactivity

Section 1

1. Californium is a radioactive element with a half-life of 44 minutes. If a sample of californium is giving a count rate of 500 counts per minute, what will be the count rate after (a) 44 minutes, (b) 88 minutes, (c) 132 minutes?

2. A scientist succeeds in isolating 0.4 g of the radioactive isotope, einsteinium. It has a half-life of 25 minutes. What mass of einsteinium will remain after 50 minutes, and what mass after 100 minutes?

3. A sample of nobelium, a radioactive isotope with a half-life of 3 seconds, has a count rate of 160 counts per minute. How long will it take to decrease to 20 counts per minute?

4. Plutonium has a half-life of 24 360 years. A sample of plutonium is decaying at a rate of 200 counts per minute. How long will it take to decrease to 50 counts per minute?

5. In the following equations for nuclear reactions, there are some spaces. Supply the missing mass numbers and atomic numbers.

   (a) $^{210}_{82}Pb \rightarrow \quad Bi + ^{0}_{-1}e$

   (b) $^{210}_{84}Po \rightarrow \quad Pb + ^{4}_{2}He$

   (c) $^{27}_{13}Al + ^{1}_{1}H \rightarrow ^{4}_{2}He + \quad Mg$

   (d) $\quad X + ^{4}_{2}He \rightarrow ^{17}_{9}F + ^{1}_{0}n$

   (e) $^{11}_{5}B + ^{1}_{1}H \rightarrow \quad X + ^{1}_{0}n$

   (f) $^{24}_{12}Mg + ^{4}_{2}He \rightarrow \, _{14}Si + ^{1}Y$

Section 2

1. The half-life of carbon-14 is 5 700 years. A sample of carbon-14 gives a count rate of 84 counts per minute on a Geiger-Müller counter. How long would it take for the count rate to drop to 21 counts per minute?

2. Uranium-239 has a half-life of 24 minutes. A solution of uranium-239 nitrate gives a count rate of 8 000 c.p.m. on a Geiger-Müller counter. What will be the count rate after (a) 48 minutes, and (b) 2 hours?

3. An enriched form of a radioactive isotope was obtained in solution. The solution was placed in a Geiger-Müller counter, and the radioactive count was measured at regular time intervals. The results are given in Table 11.2.

Table 11.2

Time in minutes	Rate of decay (counts per second)
0	600
30	225
60	81
90	30
120	11

(a) Plot a graph of the count rate against time.

(b) Deduce from your graph the count rate at 50 minutes and at 100 minutes.

(c) After what length of time was the count rate (i) 200 c.p.s. and (ii) 100 c.p.s.?

(d) Calculate the half-life of the radioactive element.

(e) What would the count rate be at 144 minutes?

(f) If a solution of half the concentration had been used, what would the count rate have been after 50 minutes and after 100 minutes?

(g) What would the half-life have been, measured in this solution?

4. A sample of a radioactive isotope was put into a Geiger-Müller counter, and the count rate was measured at various times. The results are shown in Table 11.3.

Table 11.3

Time (days)	Count rate (counts per minute)
0	520
50	410
100	320
200	195
300	115
400	70

Plot a graph of count rate against time. Use the horizontal axis for time.

(a) Use the graph to find the count rate after 150 days and after 250 days.

(b) After what length of time was the count rate (i) 300 c.p.m. (ii) 150 c.p.m.?

(c) What is the half-life of the radioactive element?

5. A radioactive isotope, A, has a half-life of 30 minutes. The initial rate of decay of A is 320 counts per minute.

(a) Draw a graph of the decay of A over the first three hours.

(b) If the original sample of A had mass 64 g, what mass of A remains after three hours?

Section 3: Questions from GCE O-level Papers

1. 8 kg of plutonium, $^{239}_{94}$Pu, are released into the atmosphere in a nuclear explosion. Deduce the mass of plutonium remaining after 72 000 years given that the half life of plutonium is 24 000 years.
    (Oxford)

2. Explain briefly why the following statement is misleading: 'The half-life of strontium-90 is 28 years. So in 56 years all the radioactivity from a sample of this isotope will have disappeared.' (Oxford)

3. A sample of 'dead' wood containing the radioactive isotope carbon-14 has an activity of 15 counts per minute, whereas the same mass of wood from a tree recently cut down has an activity of 120 counts per minute. Given that the half-life of carbon-14 is 5 570 years, calculate the age of the specimen of 'dead' wood.   (Oxford)

4. The half-life of the isotope ($^{14}_{6}$C) is 5 600 years. If 20 g of this isotope decay, calculate the mass of carbon ($^{14}_{6}$C) that remains after 11 200 years.   (Welsh JEC)

5. The elements plutonium (Pu) and neptunium (Np) are formed in the atomic reactor by normal radioactive decay in accordance with the following sequence:

$$^{239}_{92}\text{U} \xrightarrow{\ \beta\text{-emission}\ } \text{Np} \xrightarrow{\hspace{2cm}} {}_{94}\text{Pu}$$

    In the appropriate spaces in the given sequence, enter the missing mass numbers, atomic number and type of radioactive emission (α- or β-).

    Plutonium is an α-emitter, and has a half-life of 24 000 years. Give the symbol of the atom formed by radioactive decay of plutonium, including the mass number and atomic number.

    If 20 g of plutonium decay, how long will it take for the mass of plutonium to become 5 g?   (Welsh JEC)

Answers to Problems

Chapter 1

Practice with Equations

1. (a) $H_2(g) + CuO(s) \rightarrow Cu(s) + H_2O(g)$
   (b) $C(s) + CO_2(g) \rightarrow 2CO(g)$
   (c) $C(s) + O_2(g) \rightarrow CO_2(g)$
   (d) $Mg(s) + H_2SO_4(aq)$
   $\rightarrow H_2(g) + MgSO_4(aq)$
   (e) $Cu(s) + Cl_2(g) \rightarrow CuCl_2(s)$
2. (a) $Ca(s) + 2H_2O(l)$
   $\rightarrow H_2(g) + Ca(OH)_2(aq)$
   (b) $2Cu(s) + O_2(g) \rightarrow 2CuO(s)$
   (c) $4Na(s) + O_2(g) \rightarrow 2Na_2O(s)$
   (d) $Fe(s) + 2HCl(aq)$
   $\rightarrow FeCl_2(aq) + H_2(g)$
   (e) $2Fe(s) + 3Cl_2(g) \rightarrow 2FeCl_3(s)$
3. (a) $Na_2O(s) + H_2O(l) \rightarrow 2NaOH(aq)$
   (b) $2KClO_3(s) \rightarrow 2KCl(s) + 3O_2(g)$
   (c) $2H_2O_2(aq) \rightarrow 2H_2O(l) + O_2(g)$
   (d) $3Fe(s) + 2O_2(g) \rightarrow Fe_3O_4(s)$
   (e) $3Mg(s) + N_2(g) \rightarrow Mg_3N_2(s)$
   (f) $4NH_3(g) + 3O_2(g)$
   $\rightarrow 2N_2(g) + 6H_2O(g)$
   (g) $3Fe(s) + 4H_2O(g)$
   $\rightarrow Fe_3O_4(s) + 4H_2(g)$
   (h) $2H_2S(g) + 3O_2(g)$
   $\rightarrow 2H_2O(g) + 2SO_2(g)$
   (i) $2H_2S(g) + SO_2(g)$
   $\rightarrow 2H_2O(l) + 3S(s)$

Chapter 2

Problems on Relative Formula Mass

64	40	101
84	278	95
148	99	161
98	63	246
136	685	142
106	74	123.5
159.5	162	249.5
342	286	278

Problems on Percentage Composition

Section 1

1. Mg = 60%   O = 40%
2. Ca = 40%   C = 12%
   O = 48%
3. K = 39%   H = 1%
   C = 12%   O = 48%
4. (a) N = 46.7%   O = 53.3%
   (b) H = 5%   F = 95%
   (c) Be = 36%   O = 64%
   (d) Li = 46.7%   O = 53.3%
5. (a) C = 80%   H = 20%
   (b) Na = 57.5%   O = 40%
   H = 2.5%
   (c) S = 40%   O = 60%
   (d) C = 90%   H = 10%
6. (a) C = 84%   H = 16%
   (b) Mg = 72%   N = 28%
   (c) Na = 15.3%   I = 84.7%
   (d) Ca = 20%   Br = 80%

Section 2

1. (a) C = 85.7%   H = 14.3%
   (b) N = 35%   H = 5%
   O = 60%
   (c) Fe = 62.2%   O = 35.6%
   H = 2.2%
   (d) C = 26.7%   H = 2.2%
   O = 71.1%
2. (a) Fe = 28%   S = 24%
   O = 48%
   (b) 40.5%   (c) 67.5%   (d) 46.7%
3. (a) C = 60%   H = 13%
   O = 27%
   (b) C = 40%   H = 6.7%
   O = 53.3%
   (c) C = 40%   H = 6.7%
   O = 53.3%
   (d) Al = 36%   S = 64%

Chapter 3

Problems on the Mole

Section 1

1. (a) 23 g   (b) 24 g   (c) 207 g
2. (a) 13.7 g   (b) 5.2 g   (c) 11.9 g
3. (a) 254 g   (b) 216 g   (c) 54 g
    (d) 402 g
4. (a) 27 g   (b) 8 g   (c) 6 g
    (d) 10 g   (e) 5 g
5. (a) 2.0   (b) 0.05   (c) 0.75
    (d) 0.50   (e) 0.20 mol
6. (a) 44 g   (b) 98 g   (c) 36.5 g
    (d) 40 g
7. (a) 58.5 g   (b) 28 g   (c) 508 g
    (d) 265 g   (e) 13.6 g
8. (a) 111 g;   123.5 g;   171 g;   85 g
    (b) 27.75 g;   30.87 g;   42.75 g;
         21.25 g

Section 2

1. (a) 26 g   (b) 8 g   (c) 4 g
    (d) 6 g   (e) 4 g   (f) 8 g
2. (a) 2.0   (b) 2.0   (c) 0.25
    (d) 0.10   (e) 0.25 mol
3. (a) 23 g   (b) 7 g   (c) 14 g
    (d) 8 g   (e) 16 g   (f) 32 g
4. (a) 1.0   (b) 2.0   (c) 0.33
    (d) 3.0   (e) 0.5   (f) 0.125 mol
5. (a) 2070 g   (b) 10.6 g
    (c) 25.4 g   (d) 20.0 g
    (e) 10.0 g   (f) 40.0 g
    (g) 42.0 g   (h) 13.0 g
    (i) 35.5 g   (j) 2.00 g
6. (a) 1.00   (b) 0.25
    (c) 0.50   (d) 0.20
    (e) 0.20   (f) 3.0
    (g) 0.10   (h) 2.0
7. (a) 6×10^{23} (b) 6×10^{23}
    (c) 6×10^{22} (d) 3.6×10^{24}
    (e) 1.2×10^{24} (f) 6×10^{22}
    (g) 1.5×10^{22} (h) 1.2×10^{24}
8. (a) 65 g   (b) 0.065 g
9. (a) 9.0 g   (b) 0.027 g
10. (a) 12 g   (b) 0.040 g
11. (a) 20.0 g   (b) 12.0 g
    (c) 16.25 g   (d) 115 g

Problems on Reacting Masses of Solids

Section 1

1. 40.0 g
2. 10.0 g
3. 44.0 g
4. 14.67 g
5. 32.0 g
6. 4.0 g
7. 50.0 g
8. (a) 63.5 g   (b) 12.7 g
9. 4.40 g
10. 1.0 g
11. 12.0 g
12. 2.8 g

Section 2

1. 2.40 g
2. 23.2 g
3. 1.0 g
4. 22.3 g
5. 6.6 g
6. 8.0 g
7. 71.0 g
8. 27.2 g
9. 10.64 g
10. 170 g
11. (c)
12. (d)
13. (b)

Section 3

1. (a) 2.49 g   (b) 36.0%
2. (b) 8.05 g
3. (b) H_2O
4. (b) 14.3 g   (c) 2.33 g
5. 35.0%
6. 25.0 g
7. 14.3 g
8. 320 g
9. 6.80 kg
10. 23

11. (a) 176 000 tons   (b) 500 g
12. Ammonium sulphate
13. 16.8 g $NaHCO_3$;   8.2 g Na_2CO_3
14. (e)
15. (d)
16. (b)
17. (e)
18. (a)
19. (b)

Chapter 4

Problems on Empirical Formulae and Molecular Formulae

Section 1

1. Na_2O
2. Mg_3N_2
3. Fe_3O_4
4. $HgBr_2$
5. Al_2O_3
6. $BaCl_2 \cdot 2H_2O$
7. PbO_2
8. (a) C_7H_{16} (b) Mg_3N_2
    (c) Al_2S_3 (d) $CaBr_2$
    (e) Cr_2S_3

Section 2

1. (a) SO_2 (b) SO_3
    (c) NO   (d) NO_2
    (e) CH_4 (f) CH_2
2. (a) P_2O_3 (b) NH_3
    (c) Pb_3O_4 (d) SiO_2
    (e) MnO_2 (f) N_2O_5
    (g) $CrCl_3$
3. $A = C_2H_6O$ $B = C_4H_8O_2$
    $C = C_2H_6$ $D = C_6H_6$
    $E = C_3H_6$ $F = C_2H_6O_2$
    $G = C_2H_4Cl_2$ $H = C_6H_3N_3O_6$
4. (a) Na_2O (b) Pb_3O_4
    (c) NO_2 (d) Cu_2O
    (e) $FeCl_2$ (f) $FeCl_3$
5. (a) CO_2 (b) PbO_2
    (c) $CuCl_2$ (d) MgO
    (e) Mg_3N_2 (f) $AlBr_3$

6. (a) $MgSO_4 \cdot 7H_2O$
    (b) $CuSO_4 \cdot 5H_2O$
    (c) $Cr(NO_3)_3 \cdot 9H_2O$
7. (a) $MgSO_4$ (b) $N_2H_4O_3$
    (c) C_3H_8O (d) CH_2O

Section 3

1. CH_3 (a) C_2H_6 (b)

H H
| |
H—C—C—H
| |
H H

2. $MgSO_4 \cdot 7H_2O$

3. C_4H_{10}; C_6H_6

4. CH_2 (a) C_2H_4 (b) $CH_2{=}CH_2$
5. MO; MO_2
6. $BaCl_2 \cdot 2H_2O$
7. CH
8. PCl_5
9. (a) CrO_3 (b) Cr_2O_3
    (c) $4CrO_3(s) \rightarrow 2Cr_2O_3(s) + 3O_2(g)$
10. (a) CH_2 (b) C_2H_4
     (c) $CH_2{=}CH_2$ (d) ethene
     (e) the alkenes
11. (a) $A = MoO_3$
         $2MoS_2(s) + 7O_2(g)$
                 $\rightarrow 2MoO_3(s) + 4SO_2(g)$
     (b) $B = MoO_2$
         $2MoO_3(s) + Mo(s) \rightarrow 3MoO_2(s)$
12. (a) 0.38 g   (b) 0.36 g
     (c) 7
13. CH_2
14. $AlCl_3$
15. (a) C_4H_{10}
     (b) $CH_3CH_2CH{=}CH_2$ and
          $CH_3CH{=}CHCH_3$
16. Ti_2O_3 and TiO_2
17. (a) (i)   0.001 25 mol
          (ii)   0.001 25 mol
          (iii) 0.0150 g
          (iv)   0.0010 g
          (v)   0.0010 mol
     (b) (i) C_7H_5 (ii) $C_{14}H_{10}$
18. (a) $C_2H_4Cl_2$ (b) $C_2H_4Cl_2$
19. (d)
20. UO_2

21. (c)
22. (c)
23. $CaSO_4$
24. (c)

Chapter 5

Problems on Reacting Volumes of Gases at Standard Temperature and Pressure

Section 1

1. 22.4 dm³
2. 11.2 dm³
3. 2.24 dm³
4. 2.24 dm³
5. 250 cm³ O_2; 125 cm³ CO_2
6. 125 cm³ O_2

Section 2

1. 18.7 dm³ O_2; 18.7 dm³ CO_2
2. 22.7 g; 5.09 dm³
3. 325 g (a) 1 120 dm³ (b) 1 200 dm³
4. 448 cm³; 4.14 g
5. 112 cm³
6. 560 cm³
7. 0.41 g
8. 61.5 cm³
9. 112 dm³ O_2; 67.2 dm³ CO_2
10. 560 cm³; 1 120 cm³
11. 250 cm³
12. 1.00 dm³
13. A = 44; B = 64; C = 160; D = 16; E = 28; F = 17
14. A = C_2H_6O
15. B = $C_4H_8O_2$

Problems on Correction of Gas Volumes to Standard Temperature and Pressure

Section 1

1. 1.82 dm³
2. (c)

3. (b)
4. 400 cm³
5. (b)
6. (c)
7. (a) 182 cm³   (b) 91 cm³
   (c) 210 cm³   (d) 273 cm³
   (e) 150 cm³
8. (d)
9. (b)

Section 2

1. (a) 162 cm³   (b) 193 cm³
   (c) 68.3 cm³   (d) 790 cm³
2. (a) 24.6 dm³   (b) 37.4 dm³
   (c) 20.5 dm³   (d) 10.3 dm³
3. (a) 4.67 dm³   (b) 21.3 dm³
   (c) 2.95 dm³   (d) 37.4 dm³
4. (a) 0.114 dm³ or 114 cm³
   (b) 8.46 dm³   (c) 3.95 dm³
   (d) 79.5 dm³

Problems on Reacting Volumes of Gases

Section 3

1. 40 cm³ CO_2; 30 cm³ O_2
2. (a) 0.25 mol   (b) 44
3. 1.12 dm³
4. (a) 22.4 dm³   (b) 44.8 dm³
5. (a) 4 mol   (b) 8 mol
6. (c); (a)
7. (b)
8. Cl_2O
9. 1 625 cm³
10. 1.505×10^{22}; 2 500 cm³
11. 112 cm³
12. (a) 650 cm³; 400 cm³
    (b) (i) 1.505×10^{23}
        (ii) 3.01×10^{23}
13. 40 cm³ CO_2; 30 cm³ O_2
14. (a) $N/4$ (b) $4N$
    (c) $N/2$
15. (a) 15 cm³ CO_2; 50 cm³ O_2
    (b) (i) 3.01×10^{23}
        (ii) 1.20×10^{23}
16. 40; 1

17. (b) C_2H_6 (c) C_2H_4 ethene
18. (a) (i)   11.2 dm^3 (ii) 5.6 dm^3
         (iii) 2.8 dm^3
     (b) (i) $N/2$ (ii) $N/4$
         (iii) $N/8$
19. (a) 89.6 dm^3 (b) 7
     (c) 784 dm^3 (d) double
20. 35 cm^3 O_2; 10 cm^3 SO_2
21. 1.505×10^{23}; 17.75 g
22. 45 cm^3 O_2; 10 cm^3 CO_2
23. 112 cm^3
24. 20 cm^3 O_2; 20 cm^3 CO_2
25. 650 cm^3; 1.505×10^{23}
26. 70 cm^3 O_2; 40 cm^3 CO_2
27. (b)
28. 2
29. (d)
30. O_3
31. N_2O
32. (a) CH_4O (b) 32 (c) CH_4O
33. (a) 1 g H; 6 g C   (b) CH_2
     (c) 42   (d) C_3H_6

Problems involving both Masses of Solids and Volumes of Gases

Section 4

1. (a) 4.48 dm^3 (b) 1.12 dm^3
2. 4.48 dm^3
3. (a) 1.1 g   (b) 0.56 dm^3
4. (a) 120 cm^3 (b) 2.16 g
5. (a) (i)   0.05 N (ii)   0.0125 N
        (iii) 0.067 N (iv) 0.033 N
     (b) 6×10^{19}
6. 1.12 dm^3
7. 48 cm^3
8. (c)

Chapter 6

Problems on Concentration

1. (a) 0.05 mol dm^{-3}
    (b) 1.0 mol dm^{-3}
    (c) 0.25 mol dm^{-3}

    (d) 2.00 mol dm^{-3}
    (e) 0.10 mol dm^{-3}
    (f) 0.125 mol dm^{-3}
    (g) 0.25 mol dm^{-3}
    (h) 0.20 mol dm^{-3}
2. (a) 0.25 mol   (b) 0.125 mol
    (c) 0.005 mol   (d) 2.50 mol
    (e) 0.05 mol   (f) 0.025 mol
    (g) 0.37 mol   (h) 1.125 mol

Section 1

1. (a) True   (b) False
    (c) True   (d) False
    (e) True   (f) False
2. (a) True   (b) False
    (c) True   (d) False
    (e) True   (f) False
3. (a) False   (b) True
    (c) True   (d) False
    (e) False   (f) True
4. (a) True   (b) True
    (c) False   (d) False
    (e) True   (f) False
5. (a) False   (b) False
    (c) True   (d) True
    (e) True   (f) False
6. (a) True   (b) False
    (c) False   (d) False
    (e) True   (f) True

Section 2

1. 1.05 mol dm^{-3}
2. 0.55 mol dm^{-3}
3. 2.12 g Na_2CO_3; 7.88 g NaCl
     4. (a)   0.20 mol dm^{-3}
    (b) 0.143 mol dm^{-3}
    (c) 9.0 g dm^{-3}
5. 66.7 cm^3
6. 40 cm^3
7. (a) 5×10^{-3} mol   (b) 10^{-2} mol
    (c) 0.40 mol dm^{-3}
8. 50 cm^3 acid; 0.56 dm^3 CO_2
9. (a) 2.0 mol dm^{-3} (b) 400 cm^3
10. 33.3 cm^3

Section 3

1. 84.5 g
2. 200 cm^3
3. (a) 322   (b) 55.9%
   (c) 50 cm^3
4. 44.8 dm^3 CO$_2$;
   1.00 dm^3 of NaOH(aq)
5. 25 cm^3
6. 45
7. (a) 24.6 g
   (b) 0.08 mol dm^{-3} 8.48 g dm^{-3}
8. (b) (i) 25 cm^3 (ii) 50 cm^3
9. (b) 0.019   (c) 0.0095
   (d) 0.95 g   (e) 95%
10. 1.16 g
11. (a) (i)   22.5 × 10^{-3} mol
        (ii)   22.5 × 10^{-3} mol
        (iii) 0.98 mol
        (iv) 35.7 g dm^{-3}
    (b) 46 cm^3
12. (a) 0.20 g   (b) 0.10 mol
    (c) 2.24 dm^3 (d) 200 cm^3
13. (b) 0.05 mol   (c) 0.05 mol
    (d) 2.0 mol dm^{-3} (e) 50 cm^3
14. (a) 1 250 cm^3 (b) 12.5 dm^3
    (c) 1 mol
15. (a) 24 g   (b) 22.4 dm^3
    (c) 98 g or 1 dm^3 of a solution of
    concentration 1 mol dm^{-3}
    (d) 1 mol   (e) 1.12 dm^3
16. (a) 0.50 mol dm^{-3} (b) 16.7%
17. (c) 2 × 10^{-3} mol   (d) 2 × 10^{-3} mol
    (e) 0.08 mol dm^{-3}
18. (a) 2.0 mol dm^{-3} (b) 2.0 mol dm^{-3}
    (c) (i)   0.08 mol (ii)   0.04 mol
        (iii) 2.0 mol   (iv) 2
19. (a) 0.15 mol dm^{-3}
    (b) 0.125 mol dm^{-3} 7.88 g per litre
20. 0.15 mol dm^{-3}; 30 cm^3
21. (i) 6.23 g   (ii) 50 cm^3
22. (d)

Chapter 7

Problems on Electrolysis

Section 1

1. (a) 7.94 g   (b) 25.9 g

    (c) 14.9 g   (d) 20.0 g
    (e) 0.25 g   (f) 2.00 g
2. (a) 85.5 g   (b) 4.50 g
   (c) 43.8 g   (d) 7.00 g
   (e) 23.2 g   (f) 4.67 g
   (g) 25.0 g   (h) 8.00 g
3. 119 g
4. 0.52 g
5. 0.112 dm^3

Section 2

1. 965 C; 0.005 mol; 2 mol electrons
2. (a)
3. (a)
4. (b)
5. (d)
6. (b)
7. 0.25 g; 2.8 dm^3 and 8.88 g; 2.8 dm^3
8. Cd^{2+}
9. 2.4 mol; 64.8 g
10. (a)
11. (e)

Section 3

1. Cathode, 11.2 dm^3 hydrogen;
   anode, 5.6 dm^3 oxygen
3. (a) 1 Faraday;   (b) 32.5 g;
   (c) 31.75 g
4. (a) 32 g   (b) 192 000 C
   (c) Cu^{2+}
5. (a) 35.5 g chlorine
   (b) 23 g sodium
6. 0.318 g copper, 1.08 g silver
7. (b) 31.5 g   (c) 2
   (d) (i)   0.005 mol
       (ii)   0.016 g
       (iii) 0.112 dm^3
8. Cathode, 44.8 dm^3; anode, 22.4 dm^3
9. 0.5 mol electrons
10. (a) 960   (b) 0.005
    (c) 2
11. (a) (ii)   (b) (iii)
12. (c)
13. (b)
14. (a) 3 × 10^{23} (b) 2.4 × 10^{24}
15. (c)

6. (a) 579 000 C   (b) 48 250 C
7. (a) 1.2×10^{23} (b) 10.6 g

Chapter 8

Problems on Heat of Reaction

Section 1

1. 1 380 kJ
2. 12 g
3. 85.5 g
4. 45 g
5. 4 040 kJ

Section 2

1. 715 kJ mol^{-1}
2. 58.0 kJ mol^{-1}
3. 24.6 kJ mol^{-1}
4. 73 kJ mol^{-1}
5. (d)
6. 57 kJ

Section 3

1. 50 kJ
2. (b) Acid 15 °C; Alkali 23.2 °C. Alkali is warmer by 8.2 °C.
   (c) 20 cm^3 alkali and 30 cm^3 acid (2 M)
   (d) 3 M
3. (a) H_3PO_3
   (b) 2 M
   (c) $2KOH + H_3PO_3 \rightarrow K_2HPO_3 + 2H_2O$
   (d) 84 kJ mol^{-1}
4. 7 000 cm^3 (a) exothermic
   (b) 15 600 kJ   (c) 7 050 kJ
5. (b) 1 920 kJ
6. (a) Low temperature   (b) 31 kJ
   (c) evolved   (d) 11.3 g
7. (a) Al, 3.70; Cu, 1.56; Mo, 1.04; Mg, 4.17; Pt, 0.513
   (b) 46; $ScCl_3$
8. (a) (iii) 20 litres   (iv) 100 litres
   (v) negative   (vi) 11 000 g
   (b) (i) 600 cm^3 (ii) 313 °C
9. (a) D   (b) (iii)   (c) (iii)

Chapter 9

Problems on Rate of Reaction

1. (a) B   (c) 0.24 litres
2. (a) B
   (d) (i) A, 0.01   B, 0.01   C, 0.005
   (ii) A, 0.04   B, 0.04   C, 0.01
   (e) (i) A, B, C   (ii) 0.24 litres
3. (c) (i) 600 cm^3 (ii) 19 cm^3
   (d) 25 cm^3
   (e) (i) 1 dm^3 (ii) 24 dm^3
4. (a) 120 cm^3 (b) 0.005
   (c) 0.34 g   (d) 1 min 36 sec
5. (a) at the start
   (b) 120 sec
   (c) 48 cm^3
   (d) 27 sec
   (e) 2×10^{-3}
   (f) 4 cm^3
6. (v)
7. (a) (iv)   (b) (iii)
   (c) (i)   (d) (i)
   (e) (ii)
8. (d)
9. (b) 1.6 min; 1.1 min; 0.7 min
   0.39 g min^{-1}; 0.565 g min^{-1};
   0.90 g min^{-1}
10. (a)
11. (a)

Chapter 10

Problems on Solubility

1. (a) 4 g in 100 g   (b) 76 °C
   (c) 24.5 g
2. (b) 16 g
3. (a) (i) 48 g   (ii) 37 °C
   (iii) 1.00 mol dm^{-3}
   (b) $x = 6$
4. (a) 25 g and 15 g
   (c) (i) 1.7 g   (ii) 15 g
5. (a) 720 g; 19.7 moles
   (b) 19.7 mol dm^{-3}

6. (a) (iv)   (b) (iv)
    (c) (ii)   (d) (iii)
    (e) (i)
7. (c) 29.4 mol dm^{-3} (d) 770 g dm^{-3}

    (e) $NH_3 \cdot H_2O$

Chapter 11

Problems on Radioactivity

Section 1

1. (a) 250 c.p.m.   (b) 125 c.p.m.
    (c) 62.5 c.p.m.
2. (a) 0.1 g   (b) 0.025 g
3. 9 sec
4. 48 720 years
5. (a) $^{210}_{83}Bi$ (b) $^{206}_{82}Pb$
    (c) $^{24}_{12}Mg$ (d) $^{14}_{4}N$
    (e) $^{11}_{6}X$ (f) $^{27}_{14}Si + ^{1}_{0}n$

Section 2

1. 11 400 years
2. (a) 2 000 c.p.m.   (b) 250 c.p.m.
3. (b) 120 and 20 c.p.s.
    (c) (i) 33 min   (ii) 54 min
    (d) 21 min   (e) 5 c.p.s.
    (f) 60 and 10 c.p.s.   (g) 21 min
4. (a) 250 and 150 c.p.m.
    (b) (i) 115 days   (ii) 250 days
    (c) 135 days
5. (b) 1 g

Section 3

1. 1 kg
3. 16 710 years
4. 5 g
5. $^{239}_{92}U \xrightarrow{\beta} {}^{239}_{93}Np \xrightarrow{\beta} {}^{239}_{94}Pu \xrightarrow{\alpha} {}^{235}_{92}U$;
    48 000 years

List of Approximate Relative Atomic Masses

Element	Symbol	Atomic number	Relative atomic mass	Element	Symbol	Atomic number	Relative atomic mass
Aluminium	Al	13	27	Lithium	Li	3	7
Antimony	Sb	51	122	Magnesium	Mg	12	24
Argon	Ar	18	40	Manganese	Mn	25	55
Arsenic	As	33	75	Mercury	Hg	80	201
Barium	Ba	56	137	Neon	Ne	10	20
Boron	B	5	11	Nickel	Ni	28	59
Bromine	Br	35	80	Nitrogen	N	7	14
Cadmium	Cd	48	112	Oxygen	O	8	16
Caesium	Cs	55	133	Phosphorus	P	15	31
Calcium	Ca	20	40	Platinum	Pt	78	195
Carbon	C	6	12	Potassium	K	19	39
Cerium	Ce	58	140	Rubidium	Rb	37	85.5
Chlorine	Cl	17	35.5	Silicon	Si	14	28
Chromium	Cr	24	52	Silver	Ag	47	108
Cobalt	Co	27	59	Sodium	Na	11	23
Copper	Cu	29	63.5	Strontium	Sr	38	87
Fluorine	F	9	19	Sulphur	S	16	32
Gold	Au	79	197	Tin	Sn	50	119
Helium	He	2	4	Titanium	Ti	22	48
Hydrogen	H	1	1	Tungsten	W	74	184
Iodine	I	53	127	Uranium	U	92	238
Iron	Fe	26	56	Vanadium	V	23	51
Krypton	Kr	36	84	Xenon	Xe	54	131
Lead	Pb	82	207	Zinc	Zn	30	65

Logarithms

	0	1	2	3	4	5	6	7	8	9	1	2	3	4	5	6	7	8	9
10	0000	0043	0086	0128	0170						4	8	13	17	21	25	30	34	38
						0212	0253	0294	0334	0374	4	8	12	16	20	24	28	32	36
11	0414	0453	0492	0531	0569						4	8	12	15	19	23	27	31	35
						0607	0645	0682	0719	0755	4	7	11	15	18	22	26	30	33
12	0792	0828	0864	0899	0934						4	7	11	14	18	21	25	28	32
						0969	1004	1038	1072	1106	3	7	10	14	17	20	24	27	31
13	1139	1173	1206	1239	1271						3	7	10	13	16	20	23	26	30
						1303	1335	1367	1399	1430	3	6	9	13	16	19	22	25	28
14	1461	1492	1523	1553	1584						3	6	9	12	15	18	21	24	27
						1614	1644	1673	1703	1732	3	6	9	12	15	18	21	24	27
15	1761	1790	1818	1847	1875						3	6	9	11	14	17	20	23	26
						1903	1931	1959	1987	2014	3	6	8	11	14	17	19	22	25
16	2041	2068	2095	2122	2148						3	5	8	11	13	16	19	21	24
						2175	2201	2227	2253	2279	3	5	8	10	13	16	18	21	23
17	2304	2330	2355	2380	2405						3	5	8	10	13	15	18	20	23
						2430	2455	2480	2504	2529	2	5	7	10	12	15	17	20	22
18	2553	2577	2601	2625	2648						2	5	7	10	12	14	17	19	21
						2672	2695	2718	2742	2765	2	5	7	9	12	14	16	19	21
19	2788	2810	2833	2856	2878						2	5	7	9	11	14	16	18	20
						2900	2923	2945	2967	2989	2	4	7	9	11	13	15	18	20
20	3010	3032	3054	3075	3096	3118	3139	3160	3181	3201	2	4	6	8	11	13	15	17	19
21	3222	3243	3263	3284	3304	3324	3345	3365	3385	3404	2	4	6	8	10	12	14	16	18
22	3424	3444	3464	3483	3502	3522	3541	3560	3579	3598	2	4	6	8	10	12	14	15	17
23	3617	3636	3655	3674	3692	3711	3729	3747	3766	3784	2	4	6	7	9	11	13	15	17
24	3802	3820	3838	3856	3874	3892	3909	3927	3945	3962	2	4	5	7	9	11	12	14	16
25	3979	3997	4014	4031	4048	4065	4082	4099	4116	4133	2	3	5	7	9	10	12	14	15
26	4150	4166	4183	4200	4216	4232	4249	4265	4281	4298	2	3	5	7	8	10	11	13	15
27	4314	4330	4346	4362	4378	4393	4409	4425	4440	4456	2	3	5	6	8	9	11	13	14
28	4472	4487	4502	4518	4533	4548	4564	4579	4594	4609	2	3	5	6	8	9	11	12	14
29	4624	4639	4654	4669	4683	4698	4713	4728	4742	4757	1	3	4	6	7	9	10	12	13
30	4771	4786	4800	4814	4829	4843	4857	4871	4886	4900	1	3	4	6	7	9	10	11	13
31	4914	4928	4942	4955	4969	4983	4997	5011	5024	5038	1	3	4	6	7	8	10	11	12
32	5051	5065	5079	5092	5105	5119	5132	5145	5159	5172	1	3	4	5	7	8	9	11	12
33	5185	5198	5211	5224	5237	5250	5263	5276	5289	5302	1	3	4	5	6	8	9	10	12
34	5315	5328	5340	5353	5366	5378	5391	5403	5416	5428	1	3	4	5	6	8	9	10	11
35	5441	5453	5465	5478	5490	5502	5514	5527	5539	5551	1	2	4	5	6	7	9	10	11
36	5563	5575	5587	5599	5611	5623	5635	5647	5658	5670	1	2	4	5	6	7	8	10	11
37	5682	5694	5705	5717	5729	5740	5752	5763	5775	5786	1	2	3	5	6	7	8	9	10
38	5798	5809	5821	5832	5843	5855	5866	5877	5888	5899	1	2	3	5	6	7	8	9	10
39	5911	5922	5933	5944	5955	5966	5977	5988	5999	6010	1	2	3	4	5	7	8	9	10
40	6021	6031	6042	6053	6064	6075	6085	6096	6107	6117	1	2	3	4	5	6	8	9	10
41	6128	6138	6149	6160	6170	6180	6191	6201	6212	6222	1	2	3	4	5	6	7	8	9
42	6232	6243	6253	6263	6274	6284	6294	6304	6314	6325	1	2	3	4	5	6	7	8	9
43	6335	6345	6355	6365	6375	6385	6395	6405	6415	6425	1	2	3	4	5	6	7	8	9
44	6435	6444	6454	6464	6474	6484	6493	6503	6513	6522	1	2	3	4	5	6	7	8	9
45	6532	6542	6551	6561	6571	6580	6590	6599	6609	6618	1	2	3	4	5	6	7	8	9
46	6628	6637	6646	6656	6665	6675	6684	6693	6702	6712	1	2	3	4	5	6	7	7	8
47	6721	6730	6739	6749	6758	6767	6776	6785	6794	6803	1	2	3	4	5	5	6	7	8
48	6812	6821	6830	6839	6848	6857	6866	6875	6884	6893	1	2	3	4	4	5	6	7	8
49	6902	6911	6920	6928	6937	6946	6955	6964	6972	6981	1	2	3	4	4	5	6	7	8

146

Logarithms

	0	1	2	3	4	5	6	7	8	9	1	2	3	4	5	6	7	8	9
50	6990	6998	7007	7016	7024	7033	7042	7050	7059	7067	1	2	3	3	4	5	6	7	8
51	7076	7084	7093	7101	7110	7118	7126	7135	7143	7152	1	2	3	3	4	5	6	7	8
52	7160	7168	7177	7185	7193	7202	7210	7218	7226	7235	1	2	2	3	4	5	6	7	7
53	7243	7251	7259	7267	7275	7284	7292	7300	7308	7316	1	2	2	3	4	5	6	6	7
54	7324	7332	7340	7348	7356	7364	7372	7380	7388	7396	1	2	2	3	4	5	6	6	7
55	7404	7412	7419	7427	7435	7443	7451	7459	7466	7474	1	2	2	3	4	5	5	6	7
56	7482	7490	7497	7505	7513	7520	7528	7536	7543	7551	1	2	2	3	4	5	5	6	7
57	7559	7566	7574	7582	7589	7597	7604	7612	7619	7627	1	2	2	3	4	5	5	6	7
58	7634	7642	7649	7657	7664	7672	7679	7686	7694	7701	1	1	2	3	4	4	5	6	7
59	7709	7716	7723	7731	7738	7745	7752	7760	7767	7774	1	1	2	3	4	4	5	6	7
60	7782	7789	7796	7803	7810	7818	7825	7832	7839	7846	1	1	2	3	4	4	5	6	6
61	7853	7860	7868	7875	7882	7889	7896	7903	7910	7917	1	1	2	3	4	4	5	6	6
62	7924	7931	7938	7945	7952	7959	7966	7973	7980	7987	1	1	2	3	3	4	5	6	6
63	7993	8000	8007	8014	8021	8028	8035	8041	8048	8055	1	1	2	3	3	4	5	5	6
64	8062	8069	8075	8082	8089	8096	8102	8109	8116	8122	1	1	2	3	3	4	5	5	6
65	8129	8136	8142	8149	8156	8162	8169	8176	8182	8189	1	1	2	3	3	4	5	5	6
66	8195	8202	8209	8215	8222	8228	8235	8241	8248	8254	1	1	2	3	3	4	5	5	6
67	8261	8267	8274	8280	8287	8293	8299	8306	8312	8319	1	1	2	3	3	4	5	5	6
68	8325	8331	8338	8344	8351	8357	8363	8370	8376	8382	1	1	2	3	3	4	4	5	6
69	8388	8395	8401	8407	8414	8420	8426	8432	8439	8445	1	1	2	2	3	4	4	5	6
70	8451	8457	8463	8470	8476	8482	8488	8494	8500	8506	1	1	2	2	3	4	4	5	6
71	8513	8519	8525	8531	8537	8543	8549	8555	8561	8567	1	1	2	2	3	4	4	5	5
72	8573	8579	8585	8591	8597	8603	8609	8615	8621	8627	1	1	2	2	3	4	4	5	5
73	8633	8639	8645	8651	8657	8663	8669	8675	8681	8686	1	1	2	2	3	4	4	5	5
74	8692	8698	8704	8710	8716	8722	8727	8733	8739	8745	1	1	2	2	3	3	4	5	5
75	8751	8756	8762	8768	8774	8779	8785	8791	8797	8802	1	1	2	2	3	3	4	5	5
76	8808	8814	8820	8825	8831	8837	8842	8848	8854	8859	1	1	2	2	3	3	4	5	5
77	8865	8871	8876	8882	8887	8893	8899	8904	8910	8915	1	1	2	2	3	3	4	4	5
78	8921	8927	8932	8938	8943	8949	8954	8960	8965	8971	1	1	2	2	3	3	4	4	5
79	8976	8982	8987	8993	8998	9004	9009	9015	9020	9025	1	1	2	2	3	3	4	4	5
80	9031	9036	9042	9047	9053	9058	9063	9069	9074	9079	1	1	2	2	3	3	4	4	5
81	9085	9090	9096	9101	9106	9112	9117	9122	9128	9133	1	1	2	2	3	3	4	4	5
82	9138	9143	9149	9154	9159	9165	9170	9175	9180	9186	1	1	2	2	3	3	4	4	5
83	9191	9196	9201	9206	9212	9217	9222	9227	9232	9238	1	1	2	2	3	3	4	4	5
84	9243	9248	9253	9258	9263	9269	9274	9279	9284	9289	1	1	2	2	3	3	4	4	5
85	9294	9299	9304	9309	9315	9320	9325	9330	9335	9340	1	1	2	2	3	3	4	4	5
86	9345	9350	9355	9360	9365	9370	9375	9380	9385	9390	1	1	2	2	3	3	4	4	5
87	9395	9400	9405	9410	9415	9420	9425	9430	9435	9440	0	1	1	2	2	3	3	4	4
88	9445	9450	9455	9460	9465	9469	9474	9479	9484	9489	0	1	1	2	2	3	3	4	4
89	9494	9499	9504	9509	9513	9518	9523	9528	9533	9538	0	1	1	2	2	3	3	4	4
90	9542	9547	9552	9557	9562	9566	9571	9576	9581	9586	0	1	1	2	2	3	3	4	4
91	9590	9595	9600	9605	9609	9614	9619	9624	9628	9633	0	1	1	2	2	3	3	4	4
92	9638	9643	9647	9652	9657	9661	9666	9671	9675	9680	0	1	1	2	2	3	3	4	4
93	9685	9689	9694	9699	9703	9708	9713	9717	9722	9727	0	1	1	2	2	3	3	4	4
94	9731	9736	9741	9745	9750	9754	9759	9763	9768	9773	0	1	1	2	2	3	3	4	4
95	9777	9782	9786	9791	9795	9800	9805	9809	9814	9818	0	1	1	2	2	3	3	4	4
96	9823	9827	9832	9836	9841	9845	9850	9854	9859	9863	0	1	1	2	2	3	3	4	4
97	9868	9872	9877	9881	9886	9890	9894	9899	9903	9908	0	1	1	2	2	3	3	4	4
98	9912	9917	9921	9926	9930	9934	9939	9943	9948	9952	0	1	1	2	2	3	3	4	4
99	9956	9961	9965	9969	9974	9978	9983	9987	9991	9996	0	1	1	2	2	3	3	3	4

Antilogarithms

	0	1	2	3	4	5	6	7	8	9	1	2	3	4	5	6	7	8	9
0.00	1000	1002	1005	1007	1009	1012	1014	1016	1019	1021	0	0	1	1	1	1	2	2	2
0.01	1023	1026	1028	1030	1033	1035	1038	1040	1042	1045	0	0	1	1	1	1	2	2	2
0.02	1047	1050	1052	1054	1057	1059	1062	1064	1067	1069	0	0	1	1	1	1	2	2	2
0.03	1072	1074	1076	1079	1081	1084	1086	1089	1091	1094	0	0	1	1	1	1	2	2	2
0.04	1096	1099	1102	1104	1107	1109	1112	1114	1117	1119	0	1	1	1	1	2	2	2	2
0.05	1122	1125	1127	1130	1132	1135	1138	1140	1143	1146	0	1	1	1	1	2	2	2	2
0.06	1148	1151	1153	1156	1159	1161	1164	1167	1169	1172	0	1	1	1	1	2	2	2	2
0.07	1175	1178	1180	1183	1186	1189	1191	1194	1197	1199	0	1	1	1	1	2	2	2	2
0.08	1202	1205	1208	1211	1213	1216	1219	1222	1225	1227	0	1	1	1	1	2	2	2	3
0.09	1230	1233	1236	1239	1242	1245	1247	1250	1253	1256	0	1	1	1	1	2	2	2	3
0.10	1259	1262	1265	1268	1271	1274	1276	1279	1282	1285	0	1	1	1	1	2	2	2	3
0.11	1288	1291	1294	1297	1300	1303	1306	1309	1312	1315	0	1	1	1	2	2	2	2	3
0.12	1318	1321	1324	1327	1330	1334	1337	1340	1343	1346	0	1	1	1	2	2	2	2	3
0.13	1349	1352	1355	1358	1361	1365	1368	1371	1374	1377	0	1	1	1	2	2	2	3	3
0.14	1380	1384	1387	1390	1393	1396	1400	1403	1406	1409	0	1	1	1	2	2	2	3	3
0.15	1413	1416	1419	1422	1426	1429	1432	1435	1439	1442	0	1	1	1	2	2	2	3	3
0.16	1445	1449	1452	1455	1459	1462	1466	1469	1472	1476	0	1	1	1	2	2	2	3	3
0.17	1479	1483	1486	1489	1493	1496	1500	1503	1507	1510	0	1	1	1	2	2	2	3	3
0.18	1514	1517	1521	1524	1528	1531	1535	1538	1542	1545	0	1	1	1	2	2	2	3	3
0.19	1549	1552	1556	1560	1563	1567	1570	1574	1578	1581	0	1	1	1	2	2	3	3	3
0.20	1585	1589	1592	1596	1600	1603	1607	1611	1614	1618	0	1	1	1	2	2	3	3	3
0.21	1622	1626	1629	1633	1637	1641	1644	1648	1652	1656	0	1	1	2	2	2	3	3	3
0.22	1660	1663	1667	1671	1675	1679	1683	1687	1690	1694	0	1	1	2	2	2	3	3	3
0.23	1698	1702	1706	1710	1714	1718	1722	1726	1730	1734	0	1	1	2	2	2	3	3	4
0.24	1738	1742	1746	1750	1754	1758	1762	1766	1770	1774	0	1	1	2	2	2	3	3	4
0.25	1778	1782	1786	1791	1795	1799	1803	1807	1811	1816	0	1	1	2	2	2	3	3	4
0.26	1820	1824	1828	1832	1837	1841	1845	1849	1854	1858	0	1	1	2	2	3	3	3	4
0.27	1862	1866	1871	1875	1879	1884	1888	1892	1897	1901	0	1	1	2	2	3	3	3	4
0.28	1905	1910	1914	1919	1923	1928	1932	1936	1941	1945	0	1	1	2	2	3	3	4	4
0.29	1950	1954	1959	1963	1968	1972	1977	1982	1986	1991	0	1	1	2	2	3	3	4	4
0.30	1995	2000	2004	2009	2014	2018	2023	2028	2032	2037	0	1	1	2	2	3	3	4	4
0.31	2042	2046	2051	2056	2061	2065	2070	2075	2080	2084	0	1	1	2	2	3	3	4	4
0.32	2089	2094	2099	2104	2109	2113	2118	2123	2128	2133	0	1	1	2	2	3	3	4	4
0.33	2138	2143	2148	2153	2158	2163	2168	2173	2178	2183	0	1	1	2	2	3	3	4	4
0.34	2188	2193	2198	2203	2208	2213	2218	2223	2228	2234	1	1	2	2	3	3	4	4	5
0.35	2239	2244	2249	2254	2259	2265	2270	2275	2280	2286	1	1	2	2	3	3	4	4	5
0.36	2291	2296	2301	2307	2312	2317	2323	2328	2333	2339	1	1	2	2	3	3	4	4	5
0.37	2344	2350	2355	2360	2366	2371	2377	2382	2388	2393	1	1	2	2	3	3	4	4	5
0.38	2399	2404	2410	2415	2421	2427	2432	2438	2443	2449	1	1	2	2	3	3	4	4	5
0.39	2455	2460	2466	2472	2477	2483	2489	2495	2500	2506	1	1	2	2	3	3	4	5	5
0.40	2512	2518	2523	2529	2535	2541	2547	2553	2559	2564	1	1	2	2	3	4	4	5	5
0.41	2570	2576	2582	2588	2594	2600	2606	2612	2618	2624	1	1	2	3	3	4	5	5	5
0.42	2630	2636	2642	2649	2655	2661	2667	2673	2679	2685	1	1	2	3	3	4	4	5	6
0.43	2692	2698	2704	2710	2716	2723	2729	2735	2742	2748	1	1	2	3	3	4	4	5	6
0.44	2754	2761	2767	2773	2780	2786	2793	2799	2805	2812	1	1	2	3	3	4	4	5	6
0.45	2818	2825	2831	2838	2844	2851	2858	2864	2871	2877	1	1	2	3	3	4	5	5	6
0.46	2884	2891	2897	2904	2911	2917	2924	2931	2938	2944	1	1	2	3	3	4	5	5	6
0.47	2951	2958	2965	2972	2979	2985	2992	2999	3006	3013	1	1	2	3	3	4	5	5	6
0.48	3020	3027	3034	3041	3048	3055	3062	3069	3076	3083	1	1	2	3	4	4	5	6	6
0.49	3090	3097	3105	3112	3119	3126	3133	3141	3148	3155	1	1	2	3	4	4	5	6	6

Antilogarithms

	0	1	2	3	4	5	6	7	8	9	1	2	3	4	5	6	7	8	9
0.50	3162	3170	3177	3184	3192	3199	3206	3214	3221	3228	1	1	2	3	4	4	5	6	7
0.51	3236	3243	3251	3258	3266	3273	3281	3289	3296	3304	1	2	2	3	4	5	5	6	7
0.52	3311	3319	3327	3334	3342	3350	3357	3365	3373	3381	1	2	2	3	4	5	5	6	7
0.53	3388	3396	3404	3412	3420	3428	3436	3443	3451	3459	1	2	2	3	4	5	6	6	7
0.54	3467	3475	3483	3491	3499	3508	3516	3524	3532	3540	1	2	2	3	4	5	6	6	7
0.55	3548	3556	3565	3573	3581	3589	3597	3606	3614	3622	1	2	2	3	4	5	6	7	7
0.56	3631	3639	3648	3656	3664	3673	3681	3690	3698	3707	1	2	3	3	4	5	6	7	8
0.57	3715	3724	3733	3741	3750	3758	3767	3776	3784	3793	1	2	3	3	4	5	6	7	8
0.58	3802	3811	3819	3828	3837	3846	3855	3864	3873	3882	1	2	3	4	4	5	6	7	8
0.59	3890	3899	3908	3917	3926	3936	3945	3954	3963	3972	1	2	3	4	5	5	6	7	8
0.60	3981	3990	3999	4009	4018	4027	4036	4046	4055	4064	1	2	3	4	5	6	6	7	8
0.61	4074	4083	4093	4102	4111	4121	4130	4140	4150	4159	1	2	3	4	5	6	7	8	9
0.62	4169	4178	4188	4198	4207	4217	4227	4236	4246	4256	1	2	3	4	5	6	7	8	9
0.63	4266	4276	4285	4295	4305	4315	4325	4335	4345	4355	1	2	3	4	5	6	7	8	9
0.64	4365	4375	4385	4395	4406	4416	4426	4436	4446	4457	1	2	3	4	5	6	7	8	9
0.65	4467	4477	4487	4498	4508	4519	4529	4539	4550	4560	1	2	3	4	5	6	7	8	9
0.66	4571	4581	4592	4603	4613	4624	4634	4645	4656	4667	1	2	3	4	5	6	7	9	10
0.67	4677	4688	4699	4710	4721	4732	4742	4753	4764	4775	1	2	3	4	5	7	8	9	10
0.68	4786	4797	4808	4819	4831	4842	4853	4864	4875	4887	1	2	3	4	6	7	8	9	10
0.69	4893	4909	4920	4932	4943	4955	4966	4977	4989	5000	1	2	3	5	6	7	8	9	10
0.70	5012	5023	5035	5047	5058	5070	5082	5093	5105	5117	1	2	4	5	6	7	8	9	11
0.71	5129	5140	5152	5164	5176	5188	5200	5212	5224	5236	1	2	4	5	6	7	8	10	11
0.72	5248	5260	5272	5284	5297	5309	5321	5333	5336	5358	1	2	4	5	6	7	9	10	11
0.73	5370	5383	5395	5408	5420	5433	5445	5458	5470	5483	1	3	4	5	6	8	9	10	11
0.74	5495	5508	5521	5534	5546	5559	5572	5585	5598	5610	1	3	4	5	6	8	9	10	12
0.75	5623	5636	5649	5662	5675	5689	5702	5715	5728	5741	1	3	4	5	7	8	9	10	12
0.76	5754	5768	5781	5794	5808	5821	5834	5848	5861	5875	1	3	4	5	7	8	9	11	12
0.77	5888	5902	5916	5929	5943	5957	5970	5984	5998	6012	1	3	4	5	7	8	10	11	12
0.78	6026	6039	6053	6067	6081	6095	6109	6124	6138	6152	1	3	4	6	7	8	10	11	13
0.79	6166	6180	6194	6209	6223	6237	6252	6266	6281	6295	1	3	4	6	7	9	10	11	13
0.80	6310	6324	6339	6353	6368	6383	6397	6412	6427	6442	1	3	4	6	7	9	10	12	13
0.81	6457	6471	6486	6501	6516	6531	6546	6561	6577	6592	2	3	5	6	8	9	11	12	14
0.82	6607	6622	6637	6653	6668	6683	6699	6714	6730	6745	2	3	5	6	8	9	11	12	14
0.83	6761	6776	6792	6808	6823	6839	6855	6871	6887	6902	2	3	5	6	8	9	11	13	14
0.84	6918	6934	6950	6966	6982	6998	7015	7031	7047	7063	2	3	5	6	8	10	11	13	15
0.85	7079	7096	7112	7129	7145	7161	7178	7194	7211	7228	2	3	5	7	8	10	12	13	15
0.86	7244	7261	7278	7295	7311	7328	7345	7362	7379	7396	2	3	5	7	8	10	12	13	15
0.87	7413	7430	7447	7464	7482	7499	7516	7534	7551	7568	2	3	5	7	9	10	12	14	16
0.88	7586	7603	7621	7638	7656	7674	7691	7709	7727	7745	2	4	5	7	9	11	12	14	16
0.89	7762	7780	7798	7816	7834	7852	7870	7889	7907	7925	2	4	5	7	9	11	13	14	16
0.90	7943	7962	7980	7998	8017	8035	8054	8072	8091	8110	2	4	6	7	9	11	13	15	17
0.91	8128	8147	8166	8185	8204	8222	8241	8260	8279	8299	2	4	6	8	9	11	13	15	17
0.92	8318	8337	8356	8375	8395	8414	8433	8453	8472	8492	2	4	6	8	10	12	14	15	17
0.93	8511	8531	8551	8570	8590	8610	8630	8650	8670	8690	2	4	6	8	10	12	14	16	18
0.94	8710	8730	8750	8770	8790	8810	8831	8851	8872	8892	2	4	6	8	10	12	14	16	18
0.95	8913	8933	8954	8974	8995	9016	9036	9057	9078	9099	2	4	6	8	10	12	15	17	19
0.96	9120	9141	9162	9183	9204	9226	9247	9268	9290	9311	2	4	6	8	11	13	15	17	19
0.97	9333	9354	9376	9397	9419	9441	9462	9484	9506	9528	2	4	7	9	11	13	15	17	20
0.98	9550	9572	9594	9616	9638	9661	9683	9705	9727	9750	2	4	7	9	11	13	16	18	20
0.99	9772	9795	9817	9840	9863	9886	9908	9931	9954	9977	2	5	7	9	11	14	16	18	20

Index